WHY WOMEN GROW

Also by Alice Vincent

Rootbound: Rewilding a Life
How to Grow Stuff

WHY WOMEN GROW

Stories of Soil, Sisterhood and Survival

Alice Vincent

CANONGATE

First published in Great Britain in 2023
by Canongate Books Ltd, 14 High Street, Edinburgh EH1 1TE

canongate.co.uk

1

British Library Cataloguing-in-Publication Data
A catalogue record for this book is available on
request from the British Library

ISBN 978 1 83885 543 7

Typeset in Bembo Std by Palimpsest Book Production Ltd,
Falkirk, Stirlingshire

Printed and bound in Great Britain by Clays Ltd, Elcograf S.p.A.

MIX
Paper from
responsible sources
FSC® C018072
FSC
www.fsc.org

For Anna and Heather

'Gardens are about people first and plants second.'
Katherine Swift, *The Morville Hours*

CONTENTS

INTRODUCTION

E ARLY IN THE DECEMBER BEFORE everything else happened I had my tarot read. I'd gone to a party and there was a reader there. In a dim booth of a nightclub, a few drinks in, I sat and chose, turned and listened, as she answered my question: 'What will next year be like?'

People who don't know much about tarot hope it will tell them their future, or the truth. People who know better understand that it can more often confirm what you already know but are perhaps unwilling to confront. With long fingernails, the reader turned the cards: the World, the Hanged Man, the Hermit, the Empress. She talked me through them, speaking of the potential for great change and opportunity but also a quiet threat of stasis. I needed to spend some time by myself, she advised, and turn to the women in my life for strength, wisdom and guidance if I was to achieve what I hoped for.

Months later, I thought of the Empress. She is the closest the deck has to an Earth Mother image; she represents fertility and abundance, love emerging from the ground. She also signals femininity. I thought of the Empress, and I thought of the favourite women in my life. My sister, mother, two best friends; the women I admire and cackle and plot with, whom I had been reduced to sending voice notes to over WhatsApp, most of whom I hadn't touched since winter. The pandemic had wrenched us apart and played strange tricks with time. I was trying to make sense of unexpected new things that would change my life – marriage, a new shared home – while longing for the freedom and confusion of my twenties that felt more distant than ever.

Among my friends, it felt as if lockdown had accelerated the drift from girlhood to womanhood into a silent shunt. I'd expected to celebrate the last vestiges of the lives we'd grown out of, for them to linger before they transformed. Instead, they stopped suddenly; a light switch where there should have been a twilight. The lockdowns and absence, they stole our transit time. I'd meet up with friends and find them 38 weeks pregnant and mired in a house move too complicated to explain over text. Suddenly, we'd all grown up in isolation.

What emerged in me was a loneliness that was shape-shifting, a feeling that would variously present itself as anger, melancholy and malaise. For weeks, months, my brain had felt sullen and sluggish, reacting to the confines of my home with blank refusal. I felt short of ideas and ripe with inertia. I was desperate for a connection I'd never known I needed because before, I'd always been fortunate enough to have it. And so I started to look outside, beyond the close circle with whom I was changing beyond recognition and into something less known. I sought steerage on womanhood from women I had never met.

In the middle of this stuck year, I opened a green notebook and wrote down a list of names. I listed the women I wanted to speak to – strangers, most of them – about their gardens and about their lives; women whose work had interested me. Because women have always gardened, but our stories have been buried with our work. For centuries we have learned the soil's secrets. We have ushered herbs from the ground and dried them for healing; we have braided seeds into our hair to preserve legacies even when the future looks bloody and uncertain; we have silently made the world more beautiful, too often without acknowledgement. I wanted to try and change that. I wanted to see the gardens that women made. I wanted to know what had encouraged them to go out, work the soil, plant seeds and nurture them, even when so many other responsibilities sat upon their shoulders. I wanted to know how their lives had taken them to this place, and what it brought them now they were here.

For six years I had grown things with quiet compulsion

on balconies, the first one smaller and brighter than the other. I had gardened through joy and through loneliness, through communality and through heartbreak. I had gardened to celebrate and to soothe. I had become dependent on plants, on looking at and to them, to feel balanced. After nearly three years of turning a shady strip of a balcony into a verdant, welcoming oasis, I was faced with the blank expanse of a back garden.

I have always learned about plants through their stories: how they came here, what they represent, what silent powers they hold and who they mean something to. I have made my career as a storyteller: as a journalist, I have told stories daily for more than a decade. Now, I wanted to hear – perhaps even tell – these women's stories. I wanted to learn more about what had driven these women to garden, perhaps to better understand my own need for the soil, perhaps to better understand what it was to be a woman.

At first, I approached women I encountered through social media or researching online, asking if they would meet with me in a green space of their choice – pandemic restrictions willing, us living between lockdowns at the time. But I was increasingly conscious that there were many narratives and many women that lay beyond my reach, so I made a simple online form. Along with a few basic questions – age range, location – was the one I would come to find most crucial and addictive: 'What drew you to gardening?' I shared the link online and by the next morning there were more than 500 responses.

I read through them with hunger, these notes from generous strangers. All corners of life were here, in short, no-nonsense sentences. Postnatal depression, loss, grief, migration, recovery, identity, motherhood. These women gardened to carve space out of the situations their lives had placed them in. I saw patterns emerge: lockdown was a persistent motivator; moving to a home with a garden, perhaps unsurprisingly, was another. Some women, often older, had simply taken on gardening along with other domestic duties that their male partners quietly ignored. The word 'mum' or 'mother' came up most frequently. There had been botanist grandfathers and farming fathers who had helped to usher these women towards the earth, but it was mostly other women I was reading about here.

The stories came from South London and Switzerland, New York and Newcastle upon Tyne. To read them was enthralling and bittersweet: there was no way I could speak to all 700 of the women who eventually responded; I would be doing well to manage 10 per cent. Some, though, fascinated me. I sent out a flurry of emails.

I would open my inbox to find replies, with suggested dates and addresses. I snatched days to travel across the country and sometimes the continent for a few hours of conversation. Sometimes I would leave feeling flat and frustrated; the conversations were always warm and pleasant, but some left me struggling to find answers for a question I was still trying to ascertain. Still, I couldn't shake the determination to speak to women about their relationship with the earth, to find out what drew them to the soil time and time again.

Many times, I would head home buoyed on the insight I'd been fortunate enough to hear, storing away snippets of stories that I'd come to play back weeks and months after, in my own garden or while walking down the street: glimmers of wisdom that changed how I saw the world. As with many growing things, the process took time, care and good fortune. I was often surprised to hear parts of a conversation I'd had replaying in my brain while doing something wholly unrelated.

When I wanted to know why women turned to the earth, I thought about some of the reasons. I thought about grief and retreat. I thought about motherhood and creativity. I also thought about the ground as a place of political change, of the inherent politics of what it is to be a woman, to be in a body that has been othered, dismissed and fetishised for millennia. I thought about the women who see the earth as an opportunity for progress and protest.

And so, when I was looking for women to speak to, I would develop a kind of instinct about what we might discuss. Sometimes this was due to the reasons they had given me when answering my research form; sometimes it was because of what they'd posted on social media or the stories they had already told about their relationship with the ground; sometimes it was just hunch. Quite quickly, I learned that whatever I thought I might glean from an interview – whatever reasons I had guessed had taken a woman to ground – was different from what unfolded. Stories I thought might be about heritage were about salvation; stories I imagined to be about retreat ended up being about confines. I'd meet someone expecting to talk about something straightforward and leave carrying their joys, their losses, their trauma and their learnings. When this happened, I felt the substance of these conversations as heavy and solid as stone.

I was nervous before every meeting. More than anything, I was overwhelmed by the generosity of these women, who shared their lives with a stranger, and trusted her with their stories. Often, I'd share myself too, talking through my thoughts and fears about getting married and having children more intimately with these women than I had with anybody else. Sometimes, after I switched the dictaphone off, we'd sit and share a more balanced conversation, where advice and wisdom were doled out with care – often towards me, as I sought out the experiences of women who had been through what I hadn't. These women and the conversations I had with them helped me to see my life differently, but they also helped me to see my garden differently.

Nobody I spoke to had just the one reason to garden; life is rarely that cleanly cut. They found themselves outside and among plants through a variety of forces. The more I spoke to women about their gardens, the more I saw threads between them. And that is how I have presented them here, these women, these stories and the spaces that they are drawn to. As the weeks turned into seasons, what I'd heard knotted together into a tapestry. Through it, I understood how women turn to the land in order to bloom.

1

BRIXTON

I STOOD BY THE BALCONY door and watched the light
show. Late afternoon sunshine caught in the wind and
the boughs of the trees, fighting to land on the walls. An
extension lead snaked along the skirting board; a wicker
lampshade I'd found by the bins hung from the ceiling;
the fierce pride and the late nights I'd spent in the place
hung faintly in the air. I would leave it soon. But first,
one more shadowdance across the floorboards.

This space – a once-dingy, slightly damp ex-local authority
flat – had been a new start. Three years ago, I'd taken the
single brass key from the previous occupant: a handful of
stability after 15 months of shuffling around the city. I
arrived and peeled off wallpaper and polystyrene tiles;
I nested with a frenzy for a space that was wholly my
own. A friend called it Treehouse, and as I watched the
seasons change through the trees outside, the name stuck.

It offered a sense of hard-won, surviving solitude: that this was just for me.

In those first few weeks, I told myself the balcony would wait; there was too much inside to be getting on with. But within a month, on a cold and shining Saturday in late October, I caved. If I wanted colour in the spring, I had to plant bulbs in autumn. Boxes of miniature white narcissi jostled with pasta and loo roll in my supermarket shop; I ordered a handful of others through the internet and they landed on my desk at work. For more than a year I'd lost the ground beneath my feet, but now I could make a garden.

This balcony was larger than the previous one, which had held me when everything else fell apart, and smothered in shade. New ground claimed as much by me as it was squirrels and pigeons. The balcony had been left empty. But I filled it with a strange hybrid of the tropical (bamboo, *Phlebodium*, asparagus ferns) and the country garden (foxgloves, *Erigeron*, sweet woodruff, hardy geraniums) to create an otherworldly oasis. I laid down a plastic rug, I squeezed bentwood chairs through the door, I raised and lowered the flap of the melamine table depending on the season. Barely five metres square and yet large enough to lose whole weekends in. I would grow listless on the sofa, but I would devote hours to the pots and plants on the balcony, watering and feeding and gently tugging away at old foliage. The table swelled with seedling pots, the chairs became covered with planters. Sometimes I'd look back on my work with satisfaction, after several wintry hours planting bulbs, but mostly I was taken by the doing. The balance of intuition and challenge that happened when I went out there to do just one thing and came back in an hour later, fingertips darkened with soil. The balcony would often catch me off-guard – the palest blue *Iris reticulata* open at frosty daybreak, or the generous tumble of nasturtium – but what I didn't know was that I was nurturing a space that would eventually accommodate two.

Over the course of a spring and a summer, Matt and I worked and ate at the same small table, feet and laptop cables colliding on the floor. We slept until the dawn chorus woke us. One evening in May, on the balcony, we decided to get married. An unexpected change of state happened once I'd said yes, as if we'd somehow made ourselves new. A bright kind of first-date nervousness. Perhaps it was the absurdity of it: that something as simple as asking a question and giving an answer could shape the rest of our lives. It felt fantastic and preposterous, like the world had turned briefly into jelly. We kept it secret for a couple of days before we could make the socially distant rounds of his family and find friends on our daily walks; we told my family over Zoom and watched their little pixellated faces shape into surprise all at once. How heavy, how precious, how featherlight it felt.

For most of my adulthood, I didn't think I'd marry. I

didn't picture a big white wedding as a little girl, and no longing for one appeared as I got older. I loved other people's but I couldn't ignore the strange trappings so many of my friends were nudged into during them. I went to so many weddings where the bride sat quiet as speechifying men around her praised her organisational skills – as demonstrated by the wedding we were all enjoying that she, near-single-handedly, had pulled together – that I wondered where the women I knew, those fierce and funny girls, had vanished to inside their white dresses. We were all capable of more than ordering centrepieces. But here I was, with a ring on my finger, having abstract conversations about what our wedding might be like. I loved Matt, and I wanted to marry him, but I knew that binding my life to a man would come with a compromise he would not have to carry. A wedding is something that complicates and enchants. It changes the lives of the women who seek it, whatever the reason.

•

I encountered Mary Delany in a slim academic book I found on the internet, and thought she sounded like the kind of person I'd want at a party. Aristocratic, yes, but canny. She was a fearsome polymath and a creative force. A woman of letters and pictures, she wrote, painted, built gardens and dedicated grottoes to her girlfriends. Delany was in her seventies when she started to cut foliage from tissue paper and stick the leaves to black backgrounds. This was in the late 18th century; Delany was born in 1700. At a time when botanical artists were depicting the stolen spoils from colonial expansion delicately, without any of the bloodshed of their acquisition, Delany showed off their bits: sepals and stamen, pistils and petals. Forty years earlier she'd turned up to court in a near-six-foot-wide skirt of a black dress covered in botanically accurate flowers. She'd designed the entire thing herself – every other woman wore pastels. When she died, Erasmus Darwin dedicated poetry to her.

Delany resisted things and when she couldn't, she made other ones. A sharp critic of marriage, she was nevertheless subjected to it when her family's fortunes tumbled and she was married off at 17 to a gout-riddled drunk 38 years her senior. 'I was sacrificed', she wrote of the union. She was 24 when she found him, 'quite black in the face'. Her husband had never got around to changing his will, so the inheritance Delany had been married off for wasn't destined for her, but a nephew. She spent the next two decades turning down proposals and writing to her sister about the suitors she disappointed. When she was 43, she married again, against her family's wishes, to a man named Patrick, who promised her 'the encouragement to persevere in her artistic works' – and a garden.

On Patrick Delany's land Mary built a garden of her own. She carved out secret spaces and pockets of wildness;

she designed a 'Beggar's Hut' beneath a mound, a suggestive ruffle of muff-like foliage around its oval entrance. The garden fuelled Delany's creativity, as did her marriage, even if she and Patrick were more companions than lovers. After years of dismissing the institution as an imprisonment, she'd finally found a way to make it work on her terms.

I was not like Delany: not aristocratic, not born in 1700, not reliant upon a man for security or status, albeit still living in a patriarchal society. But I got engaged after swearing off marriage, and I was only gaining access to a garden thanks to the man I was marrying. Years before, when our relationship was tentative and new, I had walked Matt up my favourite street in London. It's a long, straight road up a hill in Camberwell, flanked on both sides by teetering Georgian houses with window glass so old it bends the light. The front gardens there are good for peering at. Matt told me he'd like to get me a garden, one day. It was exactly the kind of simple, preposterous ambition he tends to make – bold and big and untethered by reality. I laughed. I still think it's the most romantic thing he's ever said.

•

I pulled the car keys out of my pocket, couched them in my hand. Time to go. Ten final footsteps across the sunlight-warmed floor. This was where I had forged my independent womanhood, where I had been fortunate enough to live alone. I thought of the dinners I'd hosted around the absent table, the words I'd written there, the bills I'd paid. I thought, a little, of the statistic that women become less happy after marriage and men more so. I thought about the invisible work that happens in heteronormative couples – the buying of birthday cards, the checking of calendars, remembering the dietary requirements of those invited round for dinner. These small weights had been falling into my lap, and I knew there would be more. To build a home – a life – with someone is to compromise with them and to share yourself. In moving, Matt and I were gaining new, shining space. But I couldn't shake the notion that I was letting go of the tethers of independence I'd held when I lived alone.

I wasn't doing anything extraordinary; many of my friends lived with their partners, had got married or engaged. I was excited for the home Matt and I would build together, but I was struggling to grapple with the realities of it. To be a woman in a committed relationship in her early thirties is to constantly dance upon a line of expectation. The question of children hangs ever heavier in the air, sinking towards the ground until it emerges on every dinner and coffee table, in every pub garden, every park one occupies. I was packing boxes when my sister texted with an offer of the stuff her boys had grown out of, because now we might have a bit more room? I told her I'd feel superstitious having it in the house. There was so much I was still working out.

It was always simpler for me to push away the subject of having kids than to look at it properly. Like marriage, it was something that for most of my twenties I just thought I wouldn't do. The planet was on fire, at the time I was in a relationship with someone who didn't want them, I was wedded to my career – perhaps it was easier, *nobler*, even, just to not. Instead, I envisaged an adulthood that was indulgent and independent. Weeknight dinner parties and holidays in the middle of the school term. I vowed to be a really excellent aunt, the fun friend, the confidante my friends' children came to when they were teenagers. And yet it wasn't entirely honest: here I was in my early thirties, and a baby seemed to have snuck into what I thought could be my future. I'd never known definitively if I wanted to have or raise children, and yet I couldn't shake off motherhood as a kind of destination.

When we viewed the flat, I thought about what it would be to raise a child there. I talked about where we would keep the buggy, if the grand front steps would become a daily annoyance with a toddler in tow. I thought about placing a cot next to the bed, a high chair in the kitchen. I imagined all of the trappings of babyhood around an invisible, faceless baby; a downy head pottering around in the kitchen, a little body curled up on the sofa. These thoughts weren't conscious so much as they began to appear to me; a whiff of a life I hadn't known I wanted.

I was scared of how motherhood would change my life.

I would be bad, I thought, stuck in the house with a baby all day. I would fret over – and possibly resent – how new responsibilities would eclipse the time I had to pursue creative work. I thought about the biological changes that happen in a new mother's body; I didn't want my brain to change, for my senses to be dulled by twilight cries.

Gardens can be female spaces, like that built by Mary Delany, but they are also associated with the children women bear. I had seen big-bellied friends make gardens and new mothers wean their infants on crops from vege-table patches as many months in the growing. I had read accounts from women who, after miscarriages and failed IVF attempts, took their angry grief out on uprooting vicious Kiftsgate roses, and planting colour in the depths of winter instead. There are gardens that have changed to accommodate children – lawns to play on, flower beds swapped out for trampolines – as well as ones that exist to hold the desire to have them. When I thought of the garden Matt and I would share, I thought of peonies and sweet peas, I thought of digging and how things will grow. I was not sure I thought of a baby.

The relationship between gardening and motherhood felt uneasy and oversimplified to me. Both were messy and wild acts of creation, deeply satisfying in ways that were difficult to see from the outside. But I knew if I had a child my relationship with the garden would change. I feared it would be exposed as an indulgence, a product of time and energy I no longer had. I imagined it clut-tering up with plastic toys. Beyond the practicalities, there was something deeper: if the garden was my space to shape, a place where I could be on my own terms, what might it mean to bring a child in?

•

The last day of July and you can smell the heat in the air, see it rippling off the tarmac. I should be painting walls but end up outside in the garden, on my knees, running my fingers over the soil. It is poorer than I had anticipated, the kind of measly parched gravel that clay dries into. I pick up wrappers from water bottles, the metal casings of tealights burned out long ago. I take a bucket and water some frazzled hydrangeas. This is our first meeting, the garden and I. My first acquaintance with the soil. Right now, on this hot, dry afternoon, I can't picture the growth to come. I can't even see how the seasons will unfurl. For now, I will sit and watch where the light falls. How lucky I am to have it. From this moment, it becomes my space: Matt does not, will not, tend to the soil. I see this plot in ways he doesn't, even in this overwhelming hurl of a beginning. I see it as ground to nourish, as growth to bear witness to. He simply sees a green space beyond the windows. I will fill it with life on my own terms.

2

SOMERSET

I T IS NOT ALWAYS EASY to grow things. I have scattered many seeds only for them to lay dormant in dry soil, or fluff and moulder in that which is too wet. Sometimes the birds peck at them, sometimes the seedlings catch a hot afternoon and crisp up, limp on the edge of the pot. Sometimes everything can be done correctly, and still nothing flowers.

I turned the garden into a to-do list. In the heat of late summer, I'd broken down the rotting raised beds. Against the dusk of the first autumn nights, I sowed seeds in plastic pots and root trainers: flowers that wouldn't bloom until the following summer, dark purple sweet peas, cornflowers, bright poppies and pale pink ones. I built a flat-pack cold frame in the living room and placed it outside where the sun shines. I filled it up with seedling pots. Sometimes, late at night, I would worry about the cold before stuffing it with packaging for insulation.

I took my grandfather's stakes – two sturdy metal pegs bound together by metres of string – and pushed them, and then some canes, into the soft earth until a curve emerged across the lawn, smoothing the corners of this tatty green rectangle, unearthing space to grow after weeks of watching where the light landed. Matt and I disagreed on the width of the beds: I wanted them larger, he wanted to preserve as much lawn as possible. There was a tacit implication there: one day, there might be a child who will want to play on the grass. We compromised, and silently I vowed to make the beds a little wider every time I edged them. The turf was cut and lifted, in neat fuzzy squares, and piled up behind the plum tree. A boot-load of perennials was placed in this uncovered earth – grasses and salvias and persicaria and *Agastache* – and for a while they bloomed before fading with the light. We built a shed in the rain. I plunged tulip bulbs into the cooling earth, imagining the fiery display that would arrive six months later. *Fritillaria, Narcissus, Nectaroscordum*, names like prayers, all buried under pewter skies. Over the course of a few weeks, I snatched lunch breaks to empty out a dozen 50-litre bags of well-rotted manure onto these beds, urging nutrients into this depleted earth. It felt like there was no end to what could be put into this soil: bare-root peonies, bare-root geraniums, their brown tendrils disappearing on trust. Our kitchen scraps went into a big green bin in the corner to rot into something unrecognisable.

12

Nothing rose from the soil in those shrinking months, and I wished it would. All that effort couldn't stop life retreating; the ground was sad and sodden. The garden felt both wild and lifeless; in either case, I was not in control of it. It was stubborn, I was pleading. I stared at it all day. Sometimes, I would hear it telling me to be patient. When the invisible baby came to mind, I wondered how I could tend to this sullen garden and a child. How could I water seeds and houseplants that already wilted, even in my dependent-free life? How much nurturing, I wondered, were women supposed to be capable of?

Babies crept into my head. When I tried on new clothes, I thought about whether they would accommodate a swelling stomach, whether they would open easily for breastfeeding. I did not know if I wanted a child or not, but perhaps this was how I was working it out. A dozen of our friends were expecting children across a handful of months, and with their news came the understanding that our lives had shifted. I was not ready for it, but I'm not sure anyone ever is.

Increasingly, I had conversations that drifted towards motherhood. Older women would talk about their grown children, reflect on what it was when they arrived. Younger, child-free women would discuss a near-future life in which they would become mothers. Those in between, often in the raw, demanding years of early motherhood, would talk about the books they read while walking their babies

13

around the park, or the unimaginably large gulf between the glowing expectation of motherhood and the shattering reality, or what it was to wait for a longed-for pregnancy or the frank reality of knowing that your body couldn't make a baby. This didn't happen only when I was speaking to women about why they gardened: motherhood cropped up over lunch with colleagues, or in the interviews I'd undertake as a journalist. I would hear about a woman's experience of postnatal depression in the middle of a photoshoot, or catch a decades-old memory of babyhood at a family lunch.

In time, the bloated tummies of my friends transformed into gummy-smiled bundles. Matt and I made the rounds delivering cards and tiny clothes. We cooed over these new lives; not just those of the hot, tight little fists and cradle-capped heads, but the living rooms overtaken by muslins and bottles. We listened to sleep-pattern maths and birth stories, and then we walked out again, into the open air, wondering if and when we might do it, too. It seemed such a seismic shift, to give over a life of week-night cocktails and lazy, spontaneous Sunday afternoons for one ruled by the small and domestic. Silently, though, I knew we could both see it on the horizon. We hadn't yet been bold enough to admit that we wanted it, but I could feel it hanging, quiet but increasingly heavy, in the air between us.

One weekday afternoon a black-and-white image landed

on my phone: a neatly curled spine of a baby, heart beating strongly inside my best friend's womb. I was so pleased for her, but I knew that she was occupying a space I couldn't understand, that this was one secret that we weren't in on together. My friend was calm and excited, learning things about her body that we had never been taught in class. Over the next six months, it sometimes felt like we were standing in different rooms made of glass – in sight of one another, but separate. Since our school days, we had traced one another's rites of passage. Now her body was going through something life-changing, and I could only watch from the sidelines.

•

I stumble upon a book called *Gaining Ground* in a forum discussion and order a copy from eBay. The debut novel by Canadian author Joan Barfoot is now out of print, but is written in first-person and amidst the recent flurry of nature memoirs, of people rejecting late-capitalist life for more earthly existences, it is sometimes difficult to remember that *Gaining Ground* is not one of this number.

The premise is simple – and startling. Abra is in her early thirties and living a seemingly pleasant life: her husband is a successful stockbroker, and with him and their two primary-school-aged children she lives in a large and beautiful house. Her main duties are to dress up for the dinner parties hosted by the wives of her husband's colleagues every few weeks. But Abra leaves and moves to a remote cabin on 70 acres of land with a haunted history, bought with her children's inheritance, leaving only a note for her husband to pick up the children from their neighbour. She teaches herself to chop wood and renovate furniture, she hangs lumpen wallpaper and sews wonky curtains. She embraces the outdoors, swims in the river, observes the birds and the moon. *Gaining Ground* is fuelled by a subtle and persistent examination of the performance of femininity. There are no clocks or mirrors in Abra's new home: the first chained her to a tyranny of time, of days partitioned out by tasks; the second to a vanity she no longer possesses. But there is also no remnant of her family life – and crucially, of her children. Years later, Abra is visited by her now-grown daughter, Katie, and she doesn't recognise her. She no longer acknowledges – barely remembers, even – that she has a daughter. The young woman interrupts Abra's gardening: she is 'thinning the lettuce, checking growth, weeding: I do this every day,' our narrator tells us, 'it is going to be a fine season'.

Abra is often in the garden. She sows seeds for self-sufficiency, but it is possible to read the crops that she raises as an alternative offspring. She describes her time in the garden as 'slow, delicate, careful work, protective work, and in the end something may come of it and something may not. It is the work itself, each movement, that is the point, that is the joy, and I see where it is all right.'

In a fit of frustration at her absent, strange mother, Katie pulls out rows of seedling beans. Abra describes her as 'crying and incoherent, and of course I had to stop her, she was killing my food'. The two women fight, one inconsolable and the other trying to protect her vital sustenance. But the incident results in a moment of rare connection between them: Abra holds her daughter until, 'finally, exhausted, dishevelled and dirty, she became quiet'. Katie is carried into the cabin, rested on the sofa and covered with a blanket. Only then does Abra return to the garden to assess the damage and replace the plants, 'giving them special care'.

Barfoot offers an idea of motherhood that remains controversial more than 40 years on: that it can be forgotten, ambivalent and harmful. That motherhood eclipses a woman's identity, that it gives her even more societal roles to fill and that they may break her. I inhale the book; I'm infatuated with it. I fall for its vivid, abstract cover, which shows a crop-haired, grim-faced woman tending to cabbages. I scour the internet for interviews with Barfoot from the late seventies, wanting to find out what motivated her to write something that still feels so radical and so contemporary. Unsatisfied, I delve into her website instead and find the invitation at its close both charming and irresistible: 'For bound copies of certain novels, contact Joan via email.'

Joan replies a few days later and quickly kills my assumptions. 'I'm not a gardener,' she writes, adding that more careful readers of *Gaining Ground* had already deduced this: 'I'm reminded that after it was published, someone pointed out that I had veggies ready for harvesting at times in the growth season when they wouldn't have been contemporaneously ready.' Unprompted, Joan adds that her mother gardened. 'I remember her at the kitchen table poring through seed catalogues in winter and placing her orders and designing the flower beds – her own act of artistry. Which is what flower gardening really is, isn't it?' she writes. 'I can picture all the work she put into making the beds graceful and lovely.' Joan is in her mid-seventies now, and says that with hindsight she 'regretfully and guiltily and admiringly' sees her mother's efforts in the garden, but 'at the time I somewhat resented them'.

'This is now ridiculous,' she continues, 'but I do have an enduring memory of the little beds of petunias she planted at each side of the porch steps. And when one day a ball got rolled into the petunias, she warned me to be more careful not to damage them. Which I somehow interpreted to mean she cared more about them than she did about me. As I say, ridiculous. And yet – I still dislike petunias.'

Joan isn't the first person to mention this tussle between the things a woman should nurture. In Jamaica Kincaid's *My Garden (Book),* her children, the home she makes with them and the one she grew up in all revolve around the demands of her garden. In one scene she pitches her plants and her children as rivals, when a delivery from a nursery – one that looks after plants, not children, although it's telling that the word is used for both – appears on the floor of her garage. 'The children complained,' she writes, 'and underneath their worry was the milk-money problem: had their mother spent all the money on plants, would they be hungry? They see the garden as the thing that stands between them and true happiness: my absolute attention.'

Nurture, whether in the form of motherhood or gardening, can't be as simple as its romanticised form has so long suggested. I can feel it, in my own swirling thoughts about having a baby, my fears about having control snatched from me by tiny, grabbing hands; in the defeat left by winter's dormancy taking hold in the garden.

•

Sometimes autumn is so bright, so fantastic, that it verges on surreality: a bingo card of cliche. Misty dawns, flaming leaves, light and crunchy and golden as demerara sugar. I'd driven through them all from London, south and west to a Somerset village tucked in a valley to see Marchelle. She was in her late thirties, and I'd enjoyed her pithy and insightful posts about her garden on Instagram. It was, I'd come to learn, her first proper garden; Marchelle had been a teenager when she'd moved to the UK from Trinidad to take up a scholarship at Cambridge. Her accent still tugged at her vowels; to me, it sounded like sunshine.

Marchelle was only the second of nearly 50 women that I would speak with over the next year. We'd not met before, although we'd exchanged emails. My curiosity felt formless, almost unmanageable: I knew that I wanted to speak with Marchelle, and other women who seemed to have interesting relationships with their garden, but even in the weeks of sorting a date and finding somewhere to stay I'd not thought about what I'd actually ask her once I arrived there.

The garden shuffles up one side of the valley, accompanied by a stream that is easier to hear than see. It was this water, Marchelle tells me, that drew her to the house. She loved the way its song changed depending on where you stand in the garden. It is mid-October, and the garden is taking its last, vivid gasp: masses of purple asters, the last of the scabious, nigella and salvias; one brave, bright purple foxglove clinging on five months after its siblings bloomed. I am a relative stranger who has turned up in her driveway, and so we tour the garden; it seems the logical thing to do.

We follow the sound of the water and the bricks of the path. Things are caught somewhere between flux and stasis: Marchelle, her husband and their two young children have not yet been in this home a year, and the ground is still offering her things that the previous custodians lost track of, in part because she is listening for them. Over the hot, quiet spring and summer she and her family explored this space. They built raised beds for vegetables, they scattered seed, they cut back bamboo and let light in for forgotten plants; she read books about tree pruning and used blades to let them breathe. A path was built up the hill to the gate for the children to walk up on the way to school. Marchelle dug old wine bottles from the earth and learned that at some point the house used to be a pub. She mulched 'like a motherfucker', she tells me, a broad and infectious grin spreading beneath her cheeks. The heavy clay soil has responded with fertility. In spite of all this work, she maintains that it feels 'like the garden was made for us'. There are all these uncanny things, little reminiscences and winks, that suggest that finally this place has found the right people to look after it, the right children to play here.

We eat apple cake at the picnic table until the chill sets in, and step into a conservatory buttressed with houseplants. 'So what is it you want, then?' she asks, with genuine interest. I fumble a bit, before eventually asking her why she gardens. What she says comes after a pause, and catches me off-guard: 'It came from wanting to be mothered now that I was a mother, and not having ready access to my mother.'

Marchelle's answer surprises me. I'd wanted to know why women turned to the earth, but I'd not imagined

I'd hear such distinct motivations. I am still figuring out what my garden could be, what my relationship to it might mean. But here, breathed out through the steam of a mug, is something clear and glittering. It makes me realise that gardening is vital, something beating and alive, and more deeply resonant than mere hobby. If Marchelle grows because of something so alive and visceral – out of a need to be mothered – what other connections are waiting to be unearthed?

When Marchelle was first pregnant, she encountered a deep desire to root herself after years of travelling around a different country for a string of medical jobs. She and her husband considered returning to Trinidad, but found it was no longer the same country that she had grown up in; the children no longer played on the street, they were sent to school with security. Instead, Marchelle turned to the English countryside, determined to make it somewhere she could call home. 'Having my babies and thinking about where to be, I started turning to nature. I don't know how else to explain it,' she says. She started going for walks in the countryside and learning to identify and forage for herbs. 'I thought, if I'm going to be living in this country, if I'm going to feel like I belong here, actually I need to do something about that. And the best way I'm going to feel like I belong here is to have a relationship with the land. I need to have a place with a garden, I need to have a place where I can grow in the soil.'

Marchelle is using her garden to forge a new place of nurture – somewhere that nurtures her, somewhere that she can grow. There aren't so much grand plans as scatter-gun planting, suck-it-and-see and a gloriously open mind – she tells me she sees freedom in her lack of traditional horticultural knowledge, rather than fear. There are beds filled with swapped specimens from local plant societies, there are others that are yet to have their future decided. 'I'm just having a *go*,' she says, with an exhale somewhere between euphoria and exasperation. What energy that is, what energy she has, in creating her own relationship with what grows around her and her family.

For Marchelle, becoming a mother was the start of wanting to make a garden, but I struggle to see the two co-existing in my own life. From this outsider's perspective, motherhood seems unimaginably challenging, childbearing a life force that, for all the joy and satisfaction it offers, nevertheless holds women's time and dominates their brain chemistry immeasurably. I can't separate the notion of having a child with that of some loss of my freedom – creatively, practically, intellectually. Our society underestimates women in general but it definitely under-estimates women of childbearing age; we're yet to see a government that understands the necessity of affordable childcare. In this, perhaps, the patriarchal notion of nature – what I prefer to consider the 'outside world', as arguably humans are a part of nature too – as a mother makes most sense. Both are overlooked, underestimated and dismissed by male humankind.

Something, though, connects women's perceived roles as mothers and our custodianship of the earth. In the spreadsheet of answers from the form I'd sent out when embarking on my mission, it was interesting to note how persistently maternity cropped up. Women had taken to gardening because it gave them a reason to be outside with their children; others saw the garden as a rare example of a space that was solely theirs after they had children, or offered a vital balm to the challenges of early-years parenting. When I asked, 'What drew you to gardening?' one reply read: 'spending more time at home during maternity leave 16 years ago'. I was intrigued by its brevity: the author was a woman named Mel; she had left her email address.

A week later and we're speaking over computer screens, having found a lunchtime while Mel's baby – her third child – naps. She lives near the coast in the corner of the country and has kind eyes. Through this strange intimacy Mel tells me about the woman she was 16 years earlier, a 21-year-old who had learned she was pregnant – the result of a brief fling with a man who lived in her block of flats. Gently, she unfolds her bold naivety: how she had been travelling in New Zealand during her first trimester, how she had uprooted herself from her friends and family to follow a job to Devon. Mel recalls her decision to become a single mother with pragmatism and undeniable fact – she was quite far along, and anyway, she remembers thinking, 'people have babies all the time'.

Mel moved to a village near the sea and rented a former council house with a large garden. 'When I was presented with that, it seemed like the most normal thing in the world to see that as a blank canvas and to want to cultivate it,' she tells me. Loneliness gripped her as her stomach swelled. She was, she says, too old to be among the teen mums the state kept a keener eye on, but too young to fit in with the women in their thirties who had sandwiched their pregnancies neatly into their professional lives. Her daughter's birth was impossible for Mel to imagine before it happened and, in the event, difficult. Melanie now has three children, but complications during her subsequent labours suggested that things had gone awry the first time around without her properly understanding what she had been through. 'It did restrict my bonding a little bit, and I think it went on to cause a bit of postnatal depression. I didn't really recognise it at the time.' She was given a questionnaire about her mental health but didn't know how to answer it honestly, or what might happen if she did. 'I remember getting home from the hospital. I put the car seat down on the floor, sat on my sofa and just looked at her and thought, "Well, what do you do with babies?" You know, she just stayed in that car seat for quite a while until she cried or needed something because I had no idea what to do with a baby, if you could play with a newborn baby, or how.'

There was the garden, though. As with child-rearing, Melanie admits she didn't know much about the garden. But she went to the library, borrowed books and dug holes. The space had been neglected and she undertook a small and domestic archaeology, pulling up plates and cutlery from the ground. By the time her daughter was weaning, there was food in the garden. 'I'd sit her on the rug while I gardened and would pick salad leaves, roll them up and give them to her to chew,' she says. With each summer, Melanie, her daughter and the garden grew; not in opposition, as it was for Abra in *Gaining Ground*, but together. 'She loved shelling peas from the pod and eating those fresh.' The failures put things in context. When carrot fly decimated her crop, she realised she was able to buy some from the supermarket. 'It made me a bit more realistic about when things don't work out.'

The mother-and-baby groups were among them; someone turned up with a two-month-old in a suit. 'I just thought, "I'm not really feeling this."' The garden scooped her up. 'I do think loneliness goes with being indoors a little bit,' Mel says. 'I don't think I ever felt lonely outside as such. In the garden, there's always some noise, wherever you are. I think it would be hard to dwell on that feeling if you're outside, but very easy if you're inside.'

Mel's words help me realise why I have been pushing

myself outside in these grim, grey days. Walking to the park in the weak dawn light of winter or snatching at the fading afternoons in the garden have been the moments of the day when I've felt least alone. It has been a trying autumn and a grisly winter, one caught between change and stasis. I have struggled to feel grounded. But here, in a conversation with a woman whose life is very different from mine, I have learned more about why I am drawn to the garden.

•

Days after her baby is due, I meet my best friend by chance in a suburban garden centre. She is swollen and uncomfortable, zipped up against the drizzle in an ankle-length black coat and fed up with being pregnant. Over the past few weeks, each time we've seen one another we've remarked that it might be the last before she becomes a mother. We held a tiny, illicit baby shower with a pastel-hued cake on a striped tablecloth and south-facing spring sunshine on the wall. Her pregnancy has happened under jumpers and coats; instead of the dinners and get-togethers we'd normally have, we've had slow, short walks and sat on benches, cradling steaming cups of tea as much as our anticipation. I mourn a little that this huge, tiny thing has unfolded while we've been so separated – that I've not been able to feel the baby kick, that the months have been gobbled up by distance.

At the garden centre, we joke that we have become our mothers, who also meet up at garden centres. I pick up pink pelargoniums, bronze fennel; tomato and potato plants sit in the trolley she pushes around. We wave goodbye and I wish her luck. This is the last time I see her before the baby comes. The next week, in the early hours of the morning, I wake up in an unfamiliar bed in a farmhouse in Cornwall, in the darkness of a new moon, and worry about her labouring. We have not spoken in days, and the silence is weighty.

Later, when the sun is just a pale coral line above the horizon and the frost still white on the grass, she messages: a flurry of orange-pink photos unfolds on my screen; fleshed-out, fluorescent-lit updates to that first mono-chrome sonogram. In the same dark hour I was awake, my best friend has had a daughter. We are both in strange new places − she is on a hospital ward, I am in a thick-walled house in a Cornish valley − and still, I feel deeply tethered to her. I pull on my clothes and quietly push open the front door, taking in the expanse of the valley and the cold gasp of the crystalline pale skies filling with birdsong. In my hand sit the photographs she has sent: this tight, new little face that I will come to adore. I know that everything is different now.

•

Gardening, for Louise, was driven by practicality first: she and her partner bought a doer-upper with a back garden 'completely entangled' by one 25-year-old buddleia. But she learned things in the process: changing or improving her surrounding environment had a direct effect on her moods; gardening was a rare activity that she could indulge in alone that would make her feel good. 'I think the nurturing side of me wants to invest in living things, but not people,' she had written in answer to my survey. 'As someone who has decided not to have children, I think it's imperative to have a more expansive and inclusive view of how women bring value using our inherent human skills redirected to other places.'

Louise greets me with the kind of smile that takes over a face, up past the cheeks, into her blue eyes. I'm flustered after sitting in traffic but she exudes calm. After guiding me round to the garden at the back of her cottage, Louise comes out with a couple of glasses of wine. In the tweaked-up contrast of late afternoon light, everything looks stark; the light catches the stud in Louise's nose. We sit on a patio that has been pushed up by the roots of a towering cherry tree; behind us lies a low brick wall and a plot of dirt. This is the former site of the enormous buddleia, where all of this started. There are signs of determined beginnings: pots of various sizes, with various things in. Around the back door snakes a gangly jasmine yet to thicken. Sweet pea seedlings grasp at its ankles.

Sometimes you can see a garden before it exists: Louise tells me her hopes for the space: roses, romance; the kind of garden that transforms a little at midsummer dusk. She speaks of *Tom's Midnight Garden*, of Sissinghurst, of *The Secret Garden*, places that reveal themselves to a trusted few, that hold secrets.

We speak extensively about the choice not to have children – how our generation is the first for whom this truly exists, and even then, delicately. I had wondered about what gardens could be to women who didn't have children, for whatever reason; I'd seen them occupy lifetimes the same way that children do, and Louise is the first person willing to speak to me about it. She explains it logically to me: that there is no training to become a parent; that it's a decision that shapes a lifetime; that, unlike a garden, mistakes made are more lasting and damaging. She describes the desire to have children as something physical that resides within us, and how, in her case, 'There's a hole where that should be,' she says, simply. 'I keep trying to go and be like, I really need to figure this out and be absolutely certain. And every time I go in there, I'm just like, there's nothing in there. So I have to take that as the thing.'

It makes me realise that while I might once have been like Louise, certain in my not wanting children, I no longer am. There is something murkier inside me. I have been wrestling with the idea of nurture because

I underestimated its complexity. I thought I could learn enough about motherhood to better understand whether I could inhabit it, as if all of my research would clarify the imaginary babies I couldn't shake. I thought I might find a clear line between what it is to raise a child and what it is to make a garden. The joys and sorrows of both are too complex and enormous to be made so small. And yet, Mel saw her garden and knew, without guidance, to grow in it, just as she knew she would have and raise her daughter. Marchelle saw her garden as a creative life force akin to the mothering she needed and offered. The two are intertwined, but with a cord far more intricately woven than I had imagined.

I work hard to resist a desire to try and stitch the conversations I am hearing together. I want to understand why I garden, I'd gone out on an elaborate quest for answers, and yet part of me is still looking for neat, clean answers when I put the question to other people. Often, we say goodbye and I am left with more questions than I arrived with. I interrogate, then doubt, the process; dismiss the point of it entirely, tell myself that gardens are just gardens; I am looking for depth where there is only soil. All these hours, all these miles. Perhaps I am racking up both in the pursuit of nothing much at all.

But then there are the stories. Ribbons of words and wisdom and memory, tangled up and weaving together into something I am still learning to comprehend. At times I feel almost overwhelmed by the weight of them, of being given them – often so casually and with such honesty – of holding them, of the trust placed in me that they will be handled carefully. Soon, my search for garden stories has become a crash course in womanhood. Through listening to these stories, I learn about how womanhood became and how it could be. I hear of strength and recovery, of rebellion and compliance, of creativity and heritage and motherhood. I find an unexpected liberation in the brutal, clean honesty that can come from a conversation with a relative stranger, the intimacy that's often too potent to exist easily between friends. The conversations are like tides, they lap upon me, leaving silt and offerings in their wake. Through them, I am able to question my own womanhood – my assumptions, my prejudices, my desires. Through questioning, I learn.

3

ENFIELD

THE KENT HILLS ARE PILLOWY with mist. The motorway is Friday-lunchtime quiet as I drive past them, destined for glasshouses where I'll see the same bluish view. Large, worn-in units at the end of a working farm, on the site of a women's open prison.

It is here, at a resettlement establishment that everyone calls a prison, that rehabilitating prisoners have been growing tropical houseplants to sell to customers and businesses, learning horticultural qualifications and business acumen along the way. I meet Kali, the non-inmate who runs the enterprise, and Kayla, its most dedicated – and first – employee. In the process, I interrupt an animated conversation about stock: a new delivery is coming in; the next greenhouse along might have to be cleared out to house it. The one they stand in front of is immaculate; rows of blue star ferns, *Pilea peperomioides*, *Ficus* and *Crassula* fill the steel staging on either side in smart ceramic pots.

Each plant has a wooden label, its Latin name inscribed in cursive ballpoint. A botanical poster hangs on the back wall; in the middle there is an antique desk covered with accountancy books, a roll of brown paper and a chip-and-pin machine. On the floor stands a massive prehistoric-looking fern, which I learn is destined for an influencer's kitchen.

Kayla talks me through the stock and the transformation. The glasshouses had been out of use until a few months earlier, filled with weeds that reached her hips. In the swell and stir of a stormy summer the cracks were re-gravelled, the panes were wiped down and heating was set up. The buildings are still a work in progress: just the week before a man came to help with drainage, as the greenhouses had flooded after a wet weekend. Kayla has recently expanded her team, bringing on Alex and Netty, two other women who help her with the increasing tide of customers. Several of them arrive while I am there: an elderly couple in a Rolls-Royce; a mother and teenage daughter who pick over money plants; a breathless woman with an expensive blow-dry who turns up with a pot-bound fern. Kayla calms her down and sees to it, no-nonsense, blond ponytail bobbing. Twenty minutes later, the plant has become two and the original pot is handed back with firm instructions to wash it before anything else goes back in.

In mere months, and without access to the internet,

she has become a houseplant expert and businesswoman, learning to propagate plants, take pH readings from the soil and understand the precise balance of glasshouse conditions. 'If it's a warm night, I'll be worrying about the temperature in here,' she told me. 'I've been known to pop down to double check the temperature.' The pride beams out of her. In a place that is inherently transitory, Kayla has created a legacy. When I visit the farm shop later, a man who works there corroborates, unprompted: 'One of those girls in the glasshouses loves it; she's here first thing in the morning, there after dark.'

After the plants, what Kayla speaks most about is her children. She has two, and they are in their early teens – 'Nine and eleven when I came in,' she adds. She talks about the pain of being away from them for nearly a year due to Covid. She tells me that, to distinguish between the different kinds of fancy pot they have on sale, she and the glasshouses' other women named them after their children.

The connection between nurturing the plants and the distance from her children, for Kayla, is simple and deep. 'We are all mothers,' she explains. 'We've been separated from our children. These plants are like our babies – we nurture them, we look after them. When they're grown, we send them out into the world. It is so, so satisfying.' This is a greenhouse filled with plants, but motherhood hangs in the humidity. It is where the care comes from, it distracts from the ache of absence. In prison, these plants become ersatz offspring, an outlet for an otherwise unsatisfied pull.

•

I was conscious that, so far in my conversations, everyone I had spoken to was pregnant, or had children, or knew they didn't want them. I'd encountered writing from women who had struggled to conceive and had gardened during that arduous process, but I hadn't spoken to anyone about it. I'd always wondered if a garden offered an alternative to having a child – some gardeners I knew, often older women, had mentioned this in passing. While my friends were falling pregnant, there were only some who spoke about the difficulty that had passed before the positive test – the trips to the doctors, the longing admissions that came out at the end of an evening. I knew that some people were unable to have the children they so wanted, but these things are sad and still something of a taboo. And so I'd spoken to a few charities, asking if they could pass on a request to their communities to speak to me if gardening or the ground had made their fertility journeys easier. Hannah's reply was straightforward; she said she would like to be involved.

I've travelled for three hours and I'm in the wrong place. There are two entrances to Oakwell Hall, and I'm at the one next to a play area, teeming with toddlers and prams. I've never met Hannah, only written to her by email, but I know this isn't right. It's a brisk walk down a country lane to find her. It's a fair, if flat-skied day, and winter's residual sog has left the moss bright and green on squat stone walls. Beyond them, spanning the banks of a stream, young wild garlic leafs up. I imagine I can smell the sap rising.

Oakwell Hall is a 20-minute drive from Leeds; a solid Elizabethan house with Brontë connections and acres of now-public parkland, it's something of a jewel among the deprived former industrial towns south of the city. Hannah was born and raised nearby, and this is her favourite place. I reach her breathless and overdressed in a scarf and boots. Hannah's tall, a light denim jacket clinging to her broad shoulders, hair scraped back above thick-framed black glasses. I apologise too much, and she diffuses my gibbering with a steady, lightweight forgiveness. A mental health worker, she's used to dealing with strangers.

Hannah started to garden a few years ago, after a breakdown that saw her nearly throw everything away, including her marriage. In the space that followed, she left the job she hated and realised that what she needed was 'a house somewhere that I could go to sit, to do gardening, to take my mind off things'. Previously, the couple had been living in flats without any outside space. 'It took me two flats,' she says, 'before I realised what I really needed was a garden.' She shows me photographs of the terraced house they rent, of the yard out the back where she has started to grow things, of the handsome German Shepherd the pair adopted after moving.

We start walking into the park as Hannah begins to calmly tell me what Oakwell Hall is to her. Her accent is a comforting skein of broad Yorkshire, and through her half tour, half storytelling, I learn that this landscaped parkland held her and a friend of hers over recent months – both of them struggling with their health and happiness. They'd sit in the paved semi-circle in the orchard, watch sticky summer sundowns; they'd walk the sprawling parks. Sometimes, when they stayed past dusk, they'd share a sense that they were no longer welcome on Oakwell Hall's grounds – that the ghosts rumoured to haunt the place were calling time on their solace. She gives me an insight into her working life, trying to arrange activities and events for those in the local community with poor mental health. The pandemic has made it hard to run their support groups: the local area is notoriously poverty-stricken; very few of her clients have internet access or a device to access it on. It's not been easy to help.

Hannah's infertility has isolated her, she tells me. She's in her early thirties and many of the friends she's known from school and university have got families. 'That's really difficult for me to watch, so I just had to cut ties a bit,' she says. 'I think they want to spend time with other families, too, and connect me with other mothers and things. It just didn't serve me at that time, it doesn't serve me now.' We reach a viewpoint over some crowning hills; a trickling stream makes its way into the desultory pond of a dry winter. In the summer, Hannah tells me, this is a mass of wildflowers; the perfect spot to watch the sun set. 'You get to see all sorts,' she says. We talk about her husband, the night out on which they met, how they enjoy winding one another up. That was eight years ago, and they've been trying to have a baby for the past six. The number winds me. I tell her it's a long time. 'Very long time,' she replies, staring out at the fields beyond.

Hannah's infertility stems from a brutal combination of long-undiagnosed chronic illness, infection and unfortunate circumstance. She tells her story without self-pity, but it still makes me think of all the female pain that is too often ignored by the medical establishment. Right at the beginning of all this, Hannah thought she had fallen pregnant: her periods stopped, she started browsing online for prams. But every test came back negative. 'When it was confirmed that I wasn't, that's when the ball started rolling. My world came crashing down,' she says, of the sticky web of diagnoses that dramatically lowered her fertility. 'From then on, it's just been month after month of not knowing.' I tell her I don't know how anyone lives with that, and she says she doesn't either.

She tells me that she always wanted to be a mum, in

good part to remedy the difficult upbringing she experienced herself. 'I guess I just wanted to nurture in a way that my child would need rather than to my mum's agenda,' she says. Hannah would want to allow her child to feel comfortable in expressing themselves, in having their own opinions. She describes herself as 'creative and practical', and it seems like she navigates the pain her infertility causes that way. The path she takes me along, through the woods and up a steep incline to a beauty spot, where we look out across the whole valley, is largely empty. It confirms my suspicions that she wouldn't want to meet at the entrance by the playpark.

Hannah tells me about the yard behind her house, how she shuffled in tubs and soil. She started to sow seeds. The previous summer, she grew 17 different types of sunflowers; they are her favourite. She rattles off the varieties, the pom-poms and the teddy bears, the sprawling yellow and sophisticated dark orange. Hannah pulls up the sleeve of her jacket, strokes the soft, pale skin of her forearm underneath, and tells me that's where she'll get another tattoo – just above the looping black marks on her wrist that she had inked for fertility. Why sunflowers, I ask. 'They're so strong and resilient,' Hannah replies. 'I feel like a sunflower sometimes. I feel like I have to battle against all the elements, all the odds, everything that comes at me. I have to get up and still move forward. I think I just connected with them on a very spiritual level; I resonate with them.' We talk about the birds, how they benefit from the seedheads, and Hannah says hers are still standing now, months after they flowered. 'I just left them. A part of me didn't want to ever let go of them.'

We walk for another hour or so, encountering woodland and sneaking into the grounds of the house. Hannah and I talk about all sorts – religion, her childhood, the jokes she and her husband crack with one another. She's self-deprecating, headstrong and funny; she's lobbied her HR department to change their maternity policy regarding miscarriage. I get the sense that she is someone who always does the best she can. Just before we leave, we sit on benches around the pond. Two women walk by, pushing buggies and holding little hands. Hannah says she finds it difficult to witness other women living so casually with what she can't have. 'They don't know what the pain is like,' she says. 'What it is to think about it the minute you wake up. It is a grief. Every month is a new grief.'

•

I can't shake my time with Hannah in the weeks that follow. Her words, her courage and her grief hang heavy on me. I mark the first day of spring by dropping sweet pea seedlings over to my mum and sister. They are five months old and strong now, slight ridges appearing on their stems, and tall enough to reach my fingertips if I lay the root ball along my forearm. I tell them about Hannah, how moved I was by her story, about how forgotten that part of the world felt. I realise, quite quickly, that I could never persist like she does; that I don't think my longing for a child extends that far or that deep. I wonder if this means I should not contemplate being a mother at all.

Perseverance is crucial in tending to a garden but it is also overlooked. A lot of fuss is made about knowledge and technique – how and when to prune the roses, when to coddle and when to encourage resilience, Latin names – but the most beautiful gardens, I think, are ones made with persistence and determination. It takes a lot of effort to turn a concrete yard into a sunflower patch, to persist in borrowing library books and coaxing growth from the stubborn earth without the reassurance that a grounding in botany can offer. Many of the women I spoke with had deployed this tenacity to the spaces they grew in, a refusal to let go or give up. It was rarely spoken about with pride, but rather just something that they did. As gardeners, we cling to the next season; we have an unshakeable belief in next year, when we will sow the seeds better and prune the wayward growth earlier, when we will try again. I think a lot about Hannah's grief, how much of it she carries around, but I also think of the deep and steady hope that must accompany it in order to keep trying.

A cold, reluctant little spring ensues. I rake the beds and marvel at how the soil has improved with mulch and a winter's drenching. I read the seed packets and ignore the demands for a 'fine tilth' and 'drills', scatter all sorts with abandon and loose limbs: cornflowers 'Blue Boy' and 'Black Ball'; *Papaver rhoeas* 'Amazing Grey'; *Calendula* 'Snow Princess' and 'Sunset'; white pom-pom poppy; *Nigella damascena* 'Albion Green Pod'. The packets soften with the soil from my hands, the way my body creases the pockets that hold them as I bend and stretch. It feels sort of mad and opulent to just stand there, throwing these tiny things about, firm in the knowledge that many of them could get lost in the wind or the mouths of birds. Perhaps some, a fraction, will stick and germinate. I hope so, I so hope so.

•

The arrival of a child brings much more besides. That huge rush of love people talk about, the weight of responsibility, the creeping tether of guilt and the levity of joy. But also loss; some children die. Some mothers leave hospitals without the babies they have carried. It's a horrible truth, awful enough that even now baby loss remains a societal taboo, spoken of in hushed tones and sideways glances. According to the UK charity Tommy's, it's estimated that a quarter of pregnancies in the UK end in loss during pregnancy or birth, with seven babies arriving stillborn every day. Seven families, dozens of broken hearts. It's a painfully frequent occurrence.

I found Fiona's story online, while the days were still short and grizzly. She was eight when she knew she wanted to be a mother and 29 when she had cervical cancer. Fiona and her partner Tim went through years of IVF and miscarriages before she entered a pregnancy that would last only seven months. Her waters broke early; she spent 10 days in hospital delaying labour, and hours after being told she could return home, Fiona writes, 'the umbilical cord dropped through me, compressing my baby's supply of oxygen and nutrients. In the six minutes it took to get to the operating theatre, nurses running and shouting, wheeling my bed through empty corridors, our baby died.'

Five years have passed since Fiona lost Willow, the name she and Tim gave their first child. She picks me up from her local station, on the outskirts of North London, smiling and business-like, her eyes behind large sunglasses. A conversation about TfL rapidly turns into the revelation that on the other six days of the week she is in St Thomas's Hospital with her premature daughter, Nancy. Fiona is a career journalist, and I'm one too; we can be direct. But her news catches me off-guard. I had mentally prepared for one delicate story, but not any others.

I'd messaged Fiona because in the wake of Willow's death she'd built a garden. 'My coping mechanisms,' she wrote at the time, 'are all about doing stuff.' We arrive at her extraordinary house, smuggled within a stone's throw of the M25. The garden is full of trees that, like Fiona's home, are several centuries old. The afternoon is sticky and the shade is welcome. 'If we're talking about Willow's garden, then I think that's probably where we should go,' Fiona suggests, pragmatically. We walk through a rumpus of ivy and ferns, caught between light and shade. It's the kind of garden that would foster a deep nostalgia in growing children – one of hiding holes and secret dens, of long stretches to run amok. Fiona leads me down behind some trees and into an enclave that is shielded with light hedging. It's a deeply feeling place, that much is instantly apparent. The planting makes the garden feel almost egg-shaped, with generous outer bedding surrounding another, rounder bed in the middle. There, a willow sculpture of a little girl stands defiantly, back curving at the waist, face staring towards the sky. Fiona sits on a fallen tree trunk while I settle on a seat to her left. Ferns and ivy blur the garden's boundaries. It's irrefutable that this is a special space, but it's not clear where Willow's garden begins and the one that houses it ends. Willow's garden, like the child it is dedicated to, is both an integral part of this family garden and undeniably separate from it.

'There's lots of meaning in the garden,' Fiona says, shortly after we sit down. 'It's got four fruit trees, and I put them in to represent the grandparents. In both of our families, we have lots of nieces and nephews, so there was this waiting for our children. I was so pleased when we were expecting Willow that the four grandparents were still here. There are two silver birches to represent me and Tim, the parents.' There are drifts of Japanese anemones and hydrangeas – plants that will bloom for Willow's birthday, on 10 September; the entrance to the garden is shrouded in white roses, to represent Fiona's Yorkshire heritage, and gorse, to represent the family's Scottish line. The garden isn't just designed to be tranquil, it's been created to hold the history and heritage of Willow's family; I don't think I've ever seen a domestic garden built with such care. 'We wanted, you know, everything about us for her to share,' Fiona explains.

It was snowdrops that offered Fiona some hope during that first tidal wave of grief. The winter after Willow's stillbirth had been brutal. 'Christmas was the hardest time in some ways because I couldn't bear to be in a family scenario without the child that I felt we had promised the family,' she says. She and Tim took themselves away, off to the coast, with a new puppy. 'The dream of bringing your baby back to your home is a really key moment. We just couldn't bear any of it. So the house stopped being a sanctuary for me. It became something else I'd failed at,' she says. 'But the garden didn't; in a way that felt like it was something I could change. It's about that love and nurture, isn't it? It's about finding receptacles for these feelings.'

During those bleak little days, Fiona gathered graph paper and pencils and drew out the design. Trees would have to be taken out, land levelled. It sounds like it was a very ambitious plan, I say, even if she'd not been weathering such grief. 'Tim also thought it was ambitious,' she agrees, nodding slightly. 'He felt that maybe we should control our ambition with it. But I wanted it to be a place where I imagined I would be able to sit, and I would be able to nurture it every year, and the time that I spent in it would be the time that I would be thinking about her because I knew life would carry on and she would only carry on in my heart. I wanted to have to spend time nurturing it because it was as if that was the time I would have spent nurturing her. I felt it needed to be big to contain what I needed to do in it.' In the raw clench of that late winter, the couple hired a digger and levelled the land. They picked up old roofing tiles from around the garden and marked out edges. They laid membrane and nourished the soil and collected rocks to make a path. 'It looked sterile. There was a lot of nurturing we did before we brought the plants in; we weren't ready for the plants for a long time,' Fiona remembers. 'So I guess it mirrored where we were in our grief.'

While Fiona was building Willow's garden, she was also beginning to tell her story. A journalist at the BBC, Fiona felt she should use her writing to 'do something' and the Corporation had never really reported on stillbirth before.

As well as sharing her own experiences, Fiona interviewed five other women whose stillborn children had galvanised them to change their lives in some way. Both the garden and the piece entwined: the piece came out a month before what would have been Willow's first birthday, and the garden was ready to host a gathering that celebrated Willow and raised money for stillbirth charities. 'There was a moment, then, to memorialise her,' Fiona says, 'and to say: we have done that year. We've lived through the seasons. We've planted this garden. There was a moment to move on.' What was weird, Fiona says, was that her son, Angus, was born a year to the day that the article was published: 'Things link up in ways that we can't understand.'

In some ways, Fiona's piece broke as much ground as she and Tim did in their garden. It was read by five million people; she says she still gets a couple of emails a week about it three years on. 'I'm particularly moved by those from older women who'd not felt able to name the babies they lost, or mark their birthdays,' she tells me. We talk about the taboos that still exist around pregnancy and what happens when it goes wrong; the superstitious silence before a 12-week scan that leaves many women to suffer from early miscarriage in secrecy; the lost babies who are never spoken about, and therefore never properly grieved. After Willow, Fiona went to visit her grandmother, Nancy's namesake, and spoke to her about her fifth of six children, who was stillborn. 'I asked her how she managed, and she said "I had to keep going because I had the other children", which, for someone as strong as my nana, suggests that she felt so bad that, you know, in some way she didn't want to carry on but she had to,' Fiona says. 'I think stillbirth can be so devastating that you wonder if there are reasons to carry on. You definitely hear that through all the stories I helped those women tell.'

Fiona's way of dealing with Willow's death – to write so boldly about her birth, to make a garden and open it up, to make a space that would hold birthday celebrations every September and Mother's Day – was questioned by some for being so public. But in doing so she enabled so many other women to speak. Willow's garden is undeniably a space that people must be invited into, but it represents something far larger: the permission to mark these enormous, tiny losses; the courage to carry on; the defiance of breaking the centuries-old societal silence around stillbirth.

What I find so poignant – so uplifting, even – about Willow's garden is its sense of longevity. 'So it's in its fifth year, now,' were Fiona's first words as we walked in. The pebbled path is still growing: every time they go to the beach, Fiona will pick up a stone for it – and in time, her living children will be old enough to do this too. The family have parties in the garden, and their friends know the space as one for reflection and quietude. Angus, toddling now, comes and 'helps' Fiona in the garden, as her youngest daughter Nancy will in due course. Nancy and Willow, she tells me, are twins – they both shared the same IVF petri dish for the first five days of their existence. She says there's comfort in the 'fact that Nancy will complete our family. Willow started it, and she completed it.'

Fiona and I end our time together as pragmatically as we began: a swift departure for the station, my running for the next train. As I sit down and catch my breath I realise that I could never be like Fiona: that I have never known a maternal instinct so strong, that I would not have the persistence to keep trying to have a baby in the face of so many doctors unveiling bad news. Her desire, her determination, to have a family is a marvel to me, something almighty and extraordinary, made of unknowable matter fundamentally different from the stuff in my veins. I admire it and I fear it slightly and I know quite, quite clearly that I do not possess it. Just as I knew after I met Hannah, I know that I do not want a child enough to harvest my eggs, or spend years in pursuit of one. Perhaps this means I should not be a mother. Perhaps to even have the option is a kind of privilege I do not deserve. How unfair and how uneasy so much of this is.

4

BARNES COMMON

B EFORE WOMEN GARDENED FOR BEAUTY, they gardened to heal. St John's wort for bedwetting children, lupins for ulcers and nettles for circulation, grown in gardens for centuries with knowledge passed down through mothers and sisters. Herb women, as they were known in Tudor times, were barely documented, for the same reasons that the domestic labour undertaken by women remains invisible today – it is not considered worth writing home about. They were also often poor, and poorly paid: while an apothecary would earn £60 per year in 1629, the women who grew the plants he made his medicines with had an annual salary of £4. As Margaret Willes points out in *The Gardens of the British Working Class*, these women's lives were removed from the romanticised portrayal of rosy-cheeked flower-sellers – drawings of them show 'women dressed in ragged clothes, trudging home after a heavy day's work'.

By the 18th century, some herb women took leases on market stalls and committed their names to history in the process – Hannah Smith, who lived in Grub Street in Finsbury, London; Mary Leech and Judith Vardey, specialists in 'Phisick Herbs' – but most records of herbs from that time are in the handwriting of men, John Gerard of *Gerard's Herbal* or Nicholas Culpeper, whose *Complete Herbal* (1652) was written in simple English rather than Latin and priced at threepence, a snip of the pounds commanded by the illustrated tomes used by apothecaries. It would take nearly 300 years for a woman to formally publish a herbal when Maud Grieve, a woman who reacted to the outbreak of the First World War by setting up a school to teach women to grow and gather herbs correctly for medicinal use, was supported and edited by Hilda Leyel, who had opened a shop selling herbal remedies in 1927 as a 47-year-old divorcee, becoming 'the public face of herbalism in the interwar years', according to garden historian Catherine Horwood in her book *Gardening Women*.

It is Grieve's book, *A Modern Herbal*, that sits on my shelf. I found it second-hand, a third edition published in 1994 but retaining the clipped formality and tightly printed text of the first, 60 years earlier. I bought the book out of ignorance and ambition: I know little about herbalism, and would like to be the kind of woman who can read hedgerows and meadows like a supermarket shelf, spotting the poisonous from the preventative, understanding how to grow and prepare remedies that could calm hayfeverish eyes or uterine cramps. But I am lazy and preoccupied; the herbal sits on the shelf and occasionally I browse it in search of something to try and find in Holland & Barrett. I know this is not the answer; that to understand plants as remedies is a far more thorough and holistic practice. It's something I imagine I'll take up when I'm older.

But a garden doesn't have to contain herbs to be considered healing. The first woman I spoke to when I wanted to explore why women grew was Hazel. We went out into her garden. Raindrops clung on grass stems and swaying Japanese anemone petals and trapped the new sun. The air was thick with petrichor and roses; a concrete Willy Guhl planter spilled over with *Erigeron*. Back inside, Hazel tucked one flare-jeaned ankle beneath her and talked me through a career that had spanned music production and running a successful vintage clothing business. She spoke swiftly, broke into barks of laughter and dived into whispered confidences. She told me about the rare cancer diagnosis she'd received when she was 27; how she and her father had burned a letter that outlined her chances of survival as a statistic and of the friends who looked after her when she endured chemotherapy and radio-therapy for 18 months. Before the diagnosis, Hazel had moved into her own flat, and there was no chance of cancer making her leave. 'I think I would have felt so much worse if I was in my childhood bedroom with this,' she told me. 'It was so nice to be independent and have my own place.'

When chemotherapy left her tired but restless, Hazel turned to the flat's tiny garden. She painted the fence and laid gravel, she encouraged clematis and passion flowers up the boundaries and planted hydrangeas. 'I always found something to do, even though it was so small,' Hazel remembered. She didn't know how to garden, but, she explained, she was fearless: 'I was like, "What am I fearful of?" Most people are fearful of death. What's the ultimate fear? Dying. Well, I've gone through that.' Tending to the garden offered Hazel something life-changing in the midst of her treatment. 'I look back on that time really fondly. It just gave me such peace. I'm a huge, huge believer in how much gardening can improve your mental health. And it really got me through, it really did. The sun being on your skin. When you're feeling that sick, that's all you need.' Years after recovery, Hazel took that fearlessness and her discovered need of the outdoor world to retrain as a floral designer.

I wanted to wade into the idea that gardens could be healing. People have been making gardens with healing intentions for centuries, in monasteries and hospitals. The charity Maggie's designs and builds gardens for those affected by cancer. There is deep solace to be found in places made of plants and growing things.

But I did not want to stray into this association – between healing and growing – too easily. Hazel's story had moved me deeply, but it was her sense of fearlessness that lingered with me long after I'd left her garden. I have been fortunate enough never to know what it is to face a potentially fatal illness; I never would have imagined that it would be a lack of fear that would encourage someone to garden.

In recent years 'nature' more broadly has been positioned as a healer, but I think this oversimplifies a complex relationship and one more innately grounded in us than we understand. I did not need to 'recover' from the heartbreak that made me see plants differently during my twenties – and a break-up should never be compared to the severity of life-threatening illness – but still people would ask how gardening had 'healed' me. I didn't see it that way. Spending more time observing and engaging with how plants grew allowed me to see a world I'd previously been blind to, that of the natural biological rhythms of the year. When I started to live more closely to those cycles, to grow plants and witness the changes in their lives, I was better able to understand that what happened in mine was part of something far greater.

But if we only expect nature, or our gardens, to heal us – to make us feel better, to provide a balm or solace from our daily problems – how reciprocal can our relationship with the earth really be, and how thoroughly are we understanding the problems it is facing? The more I garden and the more time I spend paying close attention to the weather and the life cycles of the plants I grow, the more aware I am of the climate crisis. I feel drought, heatwaves and strange, warm winters all the more fiercely because I see their effects in the garden. In looking blindly to gardens as spaces only to heal ourselves, we ignore the suffering we've inflicted on the earth in taking from it so relentlessly for centuries – and the opportunity we have to tread more gently. Maud Grieve kept the fading flame of herbalism alight at a crucial time, but she was arguably more concerned that people knew how to gather herbs correctly so that they would endure. She wrote in 1916: 'Unless herbs are properly gathered and made ready for the market . . . they had better be left alone till gatherers are taught how to gather, or there may soon be no herbs to collect if they are pulled up by the roots and none left for seeds.'

•

It was Maya who recommended I buy *A Modern Herbal*. We'd met through the small and twisty world of growing and gardening. She was many things: a chef, a gardener, a herbologist, someone who preferred to dig deep than talk small. Together, we'd chosen to meet and to speak on the winter solstice, the shortest day of the year and a date that started to hold new meaning for me in the latter half of my twenties. The solstice marks the moment when the punishing gloom of December pauses before receding, each day growing longer by mere minutes. An offering both clear and irrefutable: soon, the nights would be shorter.

I'd previously celebrated the solstice in small ways and alone. A few messages, perhaps, sent to those acquaintances who also appreciated it, but otherwise the day would pass with a silent notion of hope – for longer days, for the groundswell of spring, for the chance to do things better this time. But this year Maya and I had made plans. Our friendship was new and fervent, little more than a year old, but we both felt the seasons keenly. I'd asked if she wanted to meet me and told her to choose somewhere that meant something to her. And so we met at Barnes Station, at lunchtime, for a walk across the common.

We start with the plants, as many of these conversations do. Plane trees, brambles, mugwort and yarrow. Maya has trodden the winding paths – from which many routes can be taken – that cross the 100 acres of Barnes Common countless times. Her parents live around the corner, and it was this space that held her when little else could. 'I was in such a bad way that I couldn't really see anyone or do much, so getting out of bed and going for a walk was kind of all I could manage,' she tells me, years on now, elfin leather boots defying the mud beneath. Her wild curls are tucked under a hat, her long fingers in pockets. 'I got better slowly over the course of a year, and I felt like over the seasons I got to know them, what they look like in the winter when they're all bare, how it all comes out in the spring, where the shadows lie,' Maya says, as we walk along a path lined with trees. 'I felt like they were watching my depression as well. Like, "Who is this curious, sad creature? She's a bit more upbeat today." It felt like a place to get lost, and a place to get better.'

Maya moved back to London a few years ago, from rural Scotland, itself a retreat from the big city. A fast-paced career in television and then media had left her 'running on fumes'. When she fell swiftly in love with an old school friend who worked on an estate in East Lothian, it seemed like the opportunity to live the more simple life that she had been beginning to entertain in West London: growing vegetables in a tiny back garden and learning about permaculture. 'I was burnt out. Went to Scotland. We were engaged, but only after six months,' Maya explains, rattling out these key events of her life as if they were on a shopping list. And Scotland was idyllic at times. 'We'd go foraging for our breakfast, pick yellow and red raspberries,' she remembers. 'More wild garlic than you could ever deal with. All sorts of everything you could ever want.' But it was also an enormous shift. 'I was still kind of broken and shellshocked and I didn't know where I was and I didn't have my friends and family around me and

I couldn't drive. I got a driving licence very quickly, but to begin with I couldn't even go to the shops on my own; having been very independent, that suddenly all completely changed.'

Scotland was a retreat, but one that had its challenges. Maya moved into her fiancé's home, one of a run of 'tiny' cottages rented from the estate, and the pair got married. She fell for the majestic, ancient yew tree on the grounds. She loved the way the moon dominated the sky, that she could see the stars properly. Maya could walk out the back door and into the forest. Even now, she describes the soil lovingly: 'It's a bruised purple colour, just magnificent.' It was a place that 'immersed' her in nature. She began to build a garden in the little patch of untouched land of the yard outside the cottage. 'It was my first wee bit of garden. Initially it was a virgin bit of land with a hawthorn tree in it, which was pretty magical,' she says. Hawthorn has long been connected with the heart, and this meant something to her. She would grow more herbs there, enough to transform the garden into a kind of fertile womb, 'so everything was growing up and around and you could just sit in the middle'. She dragged seaweed up from the beaches and laid it on the land. Planted cuttings and swapped seeds with neighbours. From this tiny plot, Maya grew 'about 34 different varieties of various things'.

The garden grew alongside Maya's ambitions and she wanted to learn how to raise things from the soil more formally. In lieu of a traditional RHS course, she stumbled upon a diploma in herbology at the Royal Botanic Garden Edinburgh. What she expected to be an education became a community: she was one of seven women, 'all from different backgrounds, different paths'. All of the women, Maya says, had deeply personal reasons for being on the course, a story as to why they were there. Hers was to better take care of a body traditional medicine had ignored. Her own history of endocrine problems, and the medical profession's persistent dismissal of them, had directed her towards making a herb garden to benefit everyday endocrine health. The choices she had been given by the doctors – to go on the pill, to go on antidepressants, to just live with it – hadn't satisfied her. She wanted to grow remedies to heal herself instead.

Maya found a new realm of knowledge while she was studying, but what she'd escaped London for was fracturing. Bereavement, family illness and mental health struggles made her marriage difficult to maintain. She and the man she had moved to Scotland for decided to divorce; Maya returned to the city that had broken her. 'I was really worried about moving back to the city having lived in rural Scotland,' she says. 'I thought I would feel so nature-deprived; I wouldn't be able to see the moon properly, I wouldn't be able to see the stars.' Instead, Maya's return to London demonstrated how much being away had changed her. As she took herself out onto the common

she began to realise that 'it had always been here. Everything I'd been surrounded by in Scotland had seeped into
me; I'd walk down a street lined with trees and instead
of looking at the houses, I'd feel the roots underneath
me.' Her perception had been changed by living more
closely to the earth.

I've only known Maya since that shift, and it's difficult
to imagine her not in tune with her natural surroundings.
She is a rare example among the dozens of women I've
spoken to of being a friend first and an interviewee second.
I associate our friendship with the metallic smell of the
Thames near her home, the tumble of ivy over the steps
to her front door, the moments while we're walking when
she will stop and tug at something green – a little wild
fennel here, a rub of camomile there. Maya has always
brought offerings to our meetings, usually brewed up in
her kitchen or plucked from the ground. Calendula balm,
tightly sealed marmalade, lovage seedlings. After the
conversation we had on the common, we sat on a bench
by the pond and ate stewed plums from a thermos, star
anise steam spreading through the air. While we've never
spent time together outside of London, I consider our
friendship a fiercely organic one in that it has grown
almost entirely outside, and without much in the way of
influence. I have always found greater power – a deeper
connection, even – in the plants that grow in the city.
Maya has led me to a greater understanding of them; she

has shown me the generosity of the earth, that growing is better when it is reciprocal.

Our walk around Barnes Common is peppered with facts about herbs, which she points out even in this dark and dormant time of the year. We stand on an anonymous triangle that borders the common – railway station on one side, road on the other, a well-trodden pavement leading commuters from one to the next. 'This bit,' she says, 'I absolutely love. So much yarrow. If I hadn't gone to Scotland, if I hadn't studied and done the qualifications I subsequently did, it would have been another bit of scrubland. Coming back with that knowledge completely changed the way I thought of this space and I think that is so much [about] feeling connected to where we are. This has brought me as much joy, if not more, than places I've travelled to further across the world.' Then we're off again; there's mugwort to identify. 'A bitter herb. One of the sacred herbs of the druids. Used for women's issues. To me it just smells amazing, I would harvest some, make teas with it, hang it up on my door sometimes – bit of a witchy thing to do.'

So often we think of gardens as physical places. Spaces defined by hedges, by ownership, by meaning. But Maya hasn't had a garden of her own since returning to London. Even those she made in Scotland – on the estate, for her studies – were temporary. The escape she sought was as located in those dark Scottish skies as it was in her herbals

and studies; it became a space of learning and understanding that she could take anywhere to find meaning
in what was growing around her. This stretch of common
land has held Maya, and been a backdrop to her recovery
from depression, but to say it – or the medicinal plants
that grow there – healed her would be amiss. My time
with Maya suggested that the relationship between healing
and growing is an active thing, as thorny and beautiful as
the May tree that first captured her attention in Scotland.
It is more complicated than learning what plants can do
for us; we have to offer some of ourselves to the land to
reap the benefits.

•

For all the sunshine it's a cold afternoon. I've cycled 10
minutes to a smarter bit of town to visit Diana, a writer
and gardener in her eighties. A mutual friend said she was
a conversationalist with a good garden, but beyond that
I don't know what to expect. Diana opens the garage
door wrapped up in a fur-lined anorak and her hair tucked
into a baker's boy cap. It gives her the air of a survivalist.
She's wiry and keen-eyed. Beneath rimless glasses, high
cheekbones whisper of her youth.

Walking into the garden I see light before anything else.
The sun is wintry and blinding in its last blast of the day,
and it takes a moment for the greenery to shrug it off.

Tree ferns rise at foot level from a sunken alcove to my left; a path emerges in front, dark and shining black grasses and a carpet of hellebores elsewhere. The effect is transportive: this is a place of its own making, carved out from the street furniture and parked cars of the neighbourhood that surrounds it. A black cat interrupts the backlight. 'This is Bellamy, you must meet him first,' Diana says, her vowels crisp and determined.

Up the path is a bench and chair, separated by two cast iron urns, and from there the garden begins to show more of its mapping: a space divided into three rooms by box hedging and an arch of cloud-pruned, orange-barked trees. The whole thing, even in February, is verdant. Crocuses in the lawn, frogs in the fern-rimmed pond. The air fills with our voices, those of the birds, and the occasional conversational mewl from Bellamy, whose coat Diana scrunches with vigorous familiarity.

She's well-prepared for our meeting: there's a plastic wallet of documents and she double-checks that my dictaphone is working. For years, Diana interviewed gardeners herself, compiling those conversations into a book. Do I have questions, she asks, or would I prefer her to tell me a story? I ask for stories, and she tells them: of a lonely childhood in a fractious family; of a career in interior design, and a later one in writing, both stymied by perfectionism; of the tiny first garden in Chelsea that she had other people maintain. From these I pull together a sketchy understanding of a life both well-lived and somehow soaked in sadness; in her early life, Diana was jet-set and glamorous, she married twice, the second time to the heir to a chocolate fortune, she was used to the company of the great and the good but, fundamentally, was always happier alone. Among it all, this garden had been a constant, something she had lived alongside for 40 years.

From the plastic wallet Diana takes out photographs of her garden and shows me them like others might snapshots from when they were younger. There's the beginning – a large square lawn covered in snow – and a later photo, of a cultivated wilderness that looks more like something nearing its end. The middle shot is the most arresting. Taken on a sunny day, it looks like something from a late-eighties magazine; a vision of pristine, symmetrical bounty, enormous hostas and perfectly round lawns. We're sitting at the side of the garden, and the shot is taken from above and down the middle. But even from the same angle, the garden is barely recognisable now. I ask when the photograph dates from and Diana is unwavering: 'When my marriage was breaking up.'

Like Mary Delany, Diana's second marriage had granted her a garden. She was 40, hadn't gardened much before and no longer needed to work, so she spent every hour, every day in the garden. It became an obsession – she just didn't, she says, want to leave; her friends grew tired of her fascination with it. When her marriage started to falter and took her happiness with it, she found it easier to garden than face up to what was happening inside the house. Various healers came to visit. As one was leaving he commented on how beautiful her garden was; she told him it was wearing her out. 'I still hold to me what he said,' she says, her fist clenched to her chest. 'The garden will give itself back to you, but you must ask for it,' he told her. 'In other words, I must be conscious of it. I was doing all this work, making it look so amazing, and I was very unhappy. It was the marriage as much as anything.'

We get up and walk into the next room of the garden. It's wilder back here; there's a small grove of towering bamboo, trunks thick and marbled, with ferns underneath. The path wends around the back, and Diana points out the things she is most proud of – a second, secret path to tend to the beds, where she finds female frogs on their way to the pond; her compost heaps. We stand where a lawnmower blade took two of her fingers, and she regales me with how she found – and then composted – a fox shit that appeared where they had lain on the lawn. 'I like to think that part of me is gone ahead,' she tells me, gleefully. Not long after the healer told Diana to ask for the garden back, her husband left her. She was in her mid-fifties, with no children and two divorces under her belt. 'What do I do, kill myself?' she posited. Instead, she started to write the story of the garden. 'First one I did, I could

66

hardly remember how to type,' she says. 'And that's when things started to change. And I didn't die.'

I wonder if writing about the garden – this depository for her upset, an escape from an ailing relationship – also gave Diana the space to reflect on the decade of her life that she spent making it. Diana's garden had been a mirror: tight and meticulous and ultimately constricting as she devoted herself to it as a means of escape. This had become a healing space, an organic haven filled with wildlife, but it hadn't always been one.

In the wake of her marriage, Diana says she entered a 'vandal stage' in the garden, loosening the ties of the corsetry that had harnessed it and her. The stiff round lawns frayed at the edges; the paving came up. She kept taking stones away, one by one, until there was enough ground left to plant in. She made room for something new.

The garden that had once looked like a show home became something well lived-in, in the best possible way. Arguably, it has seen a lot of living. But Diana talks of dying, too. Despite being so vital to gardens, death is rarely something we associate with them. Diana smoked into her fifties and it's caught up with her; emphysema has left her with a persistent cough and a keen sense of her own mortality. She sees much of it in the garden: the willow that she planted during her first years in the garden is now gently dying back, a 'memento mori' that now houses and feeds a family of woodpeckers. 'I think of the irony, the wonder, of it aging with me,' she says. 'Of course I don't *want* to die,' she adds, unprompted. 'I want to live forever. But there's a nice saying that old gardeners can't die because they've got to see how the garden is coming on.'

It gets too cold to ignore and I gather my things. On my way out we stand in the garage, and against the sinking light Diana is silhouetted, the peak of her cap, the fluff of her coat. She asks why I'm only speaking to women. When I briefly explain, she tells me that, between her first and second marriages, she had been with a man who was coercively controlling. 'I don't know why I let it happen,' she says. After her second marriage she 'had given up', realising that she was most content in her own company. She has lived in that sprawling house by herself for 25 years, and stresses the need for the presence of a cat. In these few short minutes, as I put on my helmet, I catch a glimpse of the quietude – aloneness, perhaps – of a life. But Diana has made a place for herself to belong, where she has the company of woodpeckers, dunnocks, blue tits and robins; of frogs and Bellamy. No longer the gleaming pageantry of symmetrical perfection, but something nourishing. Diana has made the space and given herself permission to be alone. It suits her.

Once home, I look up the story Diana wrote, the one she said saved her. It doesn't exist on the internet; it was written in the early nineties and remained on paper, but I have the copy of the gardening journal on my shelf. In the garden, our conversation had skittered, thoughts had been left hanging as we moved on to other subjects away from that of her first years making the garden. Here, on the small cream pages of *Hortus*, lies her record of it from 25 years earlier. The differing accounts – mine, from having sat in this radiant, well lived-in garden; 82-year-old Diana's from recollection; middle-aged Diana's on paper – layer up like sheets of tracing paper. In between, I see how she has changed as much as the garden. Where the ambition mellowed, where life crept in and fell out, where defiance and courage carved new paths. As I read of her longing for compost, I think of the three neat wooden boxes at the back of her garden. The brown stuff inside them – all her Amazon boxes, all her teabags, all her food waste – she told me, is her greatest pride. I am sad that Diana no longer writes: she is as sharp and funny on the page as she is in person. But, she told me, quite happily, she'd been told that what journalistic brilliance she might've achieved would have wound up matched by alcoholism. Diana is a life-long perfectionist, but over the years she's learned to cede control – in the garden, of things in her life she can't change. Her garden is made of a lifetime of learning acceptance; it's possibly why it is quite so calming.

In the weeks after we meet, I hear Diana's words as I garden. My compost is a failure – anaerobic and wet and lumpen; it squelches and stinks. The bare-root rose is necrotic. Everything is small, and dark, and picked over by squirrels. When I look at the garden, I see a list of things to do rather than a place to be. In these depths of winter I see only the fierce hopes of what I want the garden to become, rather than the generous being it already is. It is so difficult to let go of this determination, this desire for control.

5

THE WELSH BORDERS

ONE BRUISED DAWN I WALK to Ruskin Park with the ambition of seeing the city beneath the streaked sky. Fog has descended by the time I get there, blocking the tallest bits out. But for the first time in days I feel I can exhale, breath hot against the air. This morning, the cardoon heads are fluffy with seeds and I tease out a few and pocket them in the hope of growing my own. I think about the chain of things, how the plants from the garden that I used to come to for refuge in such sadness, almost exactly five years earlier, have offered a future for the one I maintain in such steady contentment. I wonder if Maya sees these ghosts of herself while walking on Barnes Common. Cities hold things, although it is often only in the quiet moments that we can see them.

Diana had shown me that gardens were spaces that existed beyond their boundaries, ones that held grief and resilience; that they could reflect where we were in our

71

lives. There was a connection between the physical acts we made in our gardens – levering up stones or letting the wildflowers grow into the lawn – and how we lived. Matt and I had moved because we wanted more space to share, and we wanted a garden. Now we had it, we were defining what that space was. Like Mel, I rarely felt lonely in the garden, but I always gardened alone. Over that first autumn and winter and into the spring I tended to it solitarily: no visitors, no helpers. With every action, I was coming to shape the garden as a place of my own. Often, it felt like speaking aloud into empty air. I was retreating into the garden. Every act I committed out there felt laden with a shapeless intention to try and make it something, to carve it into a space I might recognise myself in.

The notion of gardens as places of retreat is not a novel one. Tucked behind walls and fences, they can be in-between spaces: outdoors but private; belonging both to the air and the land as well as people. We put in perimeters and pour in our efforts, all to give ourselves somewhere to sit and to be. In recent years, scientists have studied what gardeners have known for centuries: that being outside and tending to the earth is good for us. It boosts our mood, alleviates stress and calms our anxieties. When the world locked themselves in during a pandemic, those with gardens – a crucial distinction to make – turned to them to find the solace that seemed in short supply elsewhere.

It was a move that echoed hundreds of others, documented and otherwise, through history: the soldiers who grew flowers in the trenches during the First World War, the gardens that grow against the odds in Gaza, where there is barely enough water for the humans to drink. We grow in spite of horror, because it reminds us that hope can exist.

I know how plants and gardens can offer comfort in retreat because I built myself anew by forging a connection to them at a time when my life felt broken. Gardening heaved me out of heartbreak and showed me a different way to live. But that felt like an expansion of life rather than a retreat from it; a time when I emerged from an invisible chrysalis into an existence more colourful and earthy. Lockdown showed us how much we needed green space and sent many city-dwellers to the countryside in search of bigger skies and broader horizons. Having grown up in rural England and found it stultifying, I wondered how those lives would seem as the pandemic wore on; what those retreats might be when the cities woke up again.

We can, of course, retreat without moving far at all. Plants do it all the time: any period of growth, of flowering and abundance, is followed by a decline. Annuals shed their showy petals and leave seed-filled husks behind; food for the birds and another chance at new life lying in those tiny, desiccated pods. Perennials rely on retreat, descending beneath the soil until the warmth, the wet or the light – depending on the season – is plentiful enough to bring them bursting through the surface. Without winter, there is no spring: dormancy allows things to reflect and recharge; it offers a vital pause.

But that pause isn't always easy. Often, we only fall dormant when we have run out of energy to live more eagerly; when we have worked too hard, when we are grieving, when we escape because we can no longer face staying put. In *Earthed*, Rebecca Schiller's memoir about living and being diagnosed with ADHD as an adult, the *Good Life* associations of living off the land and tending to a smallholding are routinely annihilated. At one point, in the midst of a breakdown, she takes her torment out on the plot and pulls it up by its roots, 'choosing the things I loved, that were hardest to germinate, the plants I had been most looking forward to watching get taller, grow flowers and spread'. Schiller writes: 'This place was supposed to give us what we needed, what I needed. It was going to be a thing of purpose, happiness, but somehow it is wrenching us apart and pulling my hinges off.' The plot, which carries flowers and fruit and vegetables and competes for space for goats and Schiller's children, comes to reflect her mental health. When she perseveres through the medical system on a quest for a diagnosis, the plot becomes something that Schiller similarly works on with satisfaction. She learns, she writes, that 'going back to the land' – a retreat from modernity, from cosmopolitan living – is something that appears far easier than it is. Instead, the land is 'a place I didn't know or understand or even know where the understanding might begin'. The earth and the plants that emerge from it can give us plenty, but it's rarely as honeyed as we have been led to believe.

I wanted to wade into these muddy waters, to examine the narrative that life on the land would always make us feel better. Each year, I am kept quietly hostage by the return of life in the spring; winter can feel so endless, so dark, that the emergence of new shoots from the earth feels miraculous even when we technically know it is due. This retreat is so necessary in our gardens that I thought it must be necessary in our lives, too. Maya, like Schiller, had shown me that retreat – to the land and the big skies, to 'nature' beyond a city – was complex but nevertheless crucial. I wondered if, by learning to be dormant, we could live more vibrantly.

•

As April shifts into May I begin to rely upon three sepa-rate calendars: one dedicated to my personal life, another to my work and a third to the interviews I am under-taking. I spend hours in the car for less than two in a stranger's company; I book tickets and sit on sparsely timetabled, empty trains as lockdown restrictions wane.

In some cases I break the journey, with last-minute stays in cold shepherds' huts and a rain-specked picnic in the shadow of Tintern Abbey. Trains are cancelled, lunches snatched from petrol stations and scoffed in the car. I get home late and hungry. In March I speak with five women; in April it is nine; May has three conversations scheduled in and June a further six. There are text messages sent and emails exchanged, but the conversations, I am insistent, have to be in a green space of the subject's choosing; as I tell them all, I am happy to travel.

It is not always easy. I am usually meeting people for the first time only to ask them about their lives. I choose many of the subjects on instinct; that these women – whom I encounter online, or through their work, or from others' passing anecdotes – seem interesting and invite my curiosity. I usually have an idea of what they might talk to me about, of where our conversations might go, but this is nearly always challenged by what actually happens. Beyond making sure I am in the right place at the right time, I do little other preparation: my research lies in what we speak about in the moment. I want to hear these women's stories for what they are, rather than to try and fit them to expectation.

•

I knew Martha, my friend's cousin, but I'd not known she'd moved to a cabin and spent the winter there until she told me in passing. It intrigued me, the isolation this woman had put herself in. I was thinking a lot about gardens as retreats, about the luxury of having somewhere to escape to, but also about what drove women to want to leave people behind and find clarity in open spaces. It was what Maya had done, taking herself off to a place where the moon was so bright it made it feel closer. It was what Abra had done, too, albeit in Joan Barfoot's fiction. Such moves can be aspirational. So often, women – mothers, especially – speak wistfully of having a calm, quiet space where they aren't pawed at by small hands or made demands of by their partners. Where there are no dishwashers to stack, or calendars to jostle. Martha wasn't a mother and she was single, but I still found her decision radical: there is a bravery in uprooting oneself, detaching oneself from a city and a community to live in a hut in a garden on a hillside cut off by a river. A couple of years earlier I had done something similar, taken the train up to the northeast coast of the country and spent a few nights in an off-grid cabin alone, in late January. I read, poached eggs and watched the sun rise and set. It was an odd, not unpleasant kind of holiday. I could never muster the courage to turn it into a full-time affair.

I was encountering a new and uneasy loneliness that came from arriving at a stage of life I didn't know what to do with: a womanhood that I'd not read any manuals for. I was neither a mother nor completely removed from the idea of motherhood; I wasn't a wife, but my wedding was on the horizon. With each day I felt extracted from a girlhood that I had felt comfortable in, even though I had outgrown it. Renovating my own home, just for me, had felt an act of independence. But to do it with a ring on my finger – and for two people – seemed suspiciously like homemaking. I wanted to make things beautiful, I wanted to host our friends, throw parties and serve up dinners. Every time I did so, that enjoyment asked if this was a womanhood I was performing, a wifehood I was stepping into. I was scared of losing myself, my time and my creativity to the domestic labour that so many women undertake invisibly. Over the months that followed, into the squalid summer and the shrinking days beyond, I would feel strange and listless: busy with a social life and yet somehow still removed from my friends, caught in an in-between state I couldn't define. How much of this was loneliness, how much discomfort, how much boredom? These early evening hours felt the quietest; I sat with my words, with the light playing on the wall, with the clock. I strove to find grace in them. Often, I failed.

•

I open the car window to smell it: wild garlic, fresh from the rain, heady and medicinal. I am curving around the edge of the Wye Valley, on a road that holds tight close to the river that separates England from Wales in two back-to-back U-bends. The rain has been lighter here and everything is newly washed with it. Early May: unfurling ferns; luminous, spongy moss; air thick with earthy promise. After the grey of the motorway, this green abundance feels otherworldly. My destination clings to a hillside, tucked up a narrow road. I call Martha from the deserted car park, and minutes later she's running down to me in a bright yellow fisherman's jumper, arms out, beaming. We collide in a hug. Even though we don't see one another much, I hold great affection for Martha. In my mind, she's ossified as the girl she once was – who would go to a festival for a weekend with one silk dress and an anorak and leave with a summer job, a potential lover and grubby knees. I'm always a bit surprised that Martha's no longer 19.

She's been living here, in this community on a large, private plot of land, since the previous October. Seven months of near-solitude in a cabin through a tough, snowy winter, most of which has been under the strictest lockdown conditions in the world. The garden has been home to three generations of a family of botanists and ecologists who have been collecting and growing plants organically in the garden for decades, as well as a handful of other households. Martha works in the garden. In exchange, she lives in the cabin rent-free.

We are caught between winter and spring: the narcissus have died back, the hellebores are going over, the full flush of May is weeks away. But the beauty of the place is undeniable; the garden sprawls, there are ponds and stepped gardens, a kind of walled ginnel lined with moss. Martha is tall, her strides are long, and she's come to know the place well over her time here; I am whisked around the garden's nooks and crannies, unable to make sense of how it fits together. A dovecote here, an orchard there. We land at Martha's 'hobbit greenhouse' – too short for her to stand up in, although I walk in just fine. The veg plot she's inherited from the previous community members sits nearby, via a pile of well-rotted manure that conjures its own story of rural dodgy dealings. Somehow, in keeping with the illogic of the place, a kiwi vine is flourishing. 'For breakfast, every day, from December to the end of January, I was eating five kiwis,' she tells me, blue eyes wide above her freckles. 'It was so mad!'

Her patch is 'so sweet, and very weedy'. Unlike the rest of the garden, this is Martha's to tend to alone. 'This is the first garden I've ever really done by myself,' she explains. 'It can be overwhelming.' It's a place of experimentation: some beds she has dug, some she's not; there's a comfrey hedge ('Because why not? I want on-tap amazing fertiliser') and perpetual spinach and chard – gifts from her predecessors. 'We've got strawberries and raspberries, more of that phacelia stuff . . .' she rattles off. Pea sticks are trussed into wigwams with pink string. In the corner, three fancy tulips flame against the utility of it all, like flamingos lost in a farmyard.

Martha's delight with it all is infectious: she speaks in a rapid patter, switching between stories and pointing out plants. It's the start of the growing season, and she hopes to raise cut flowers and salad leaves to sell in a shop down the hill in Monmouth. She still can't believe that wild garlic literally grows on her doorstep. Sometimes she'll find offerings from other plotholders hanging on the hook by her door – a little purple-sprouting broccoli, some greens. But she has still left a city – Bristol – to come and live in a cabin away from her friends. The nearest supermarket is a 20-minute drive away. 'You can't hide from anything here,' she tells me. 'Everything that you rely on is right outside your door. If you don't sort out your firewood or if you don't do this, this and this, you're fucked, basically.' Through the trees we look at how the river bends below, the confrontation of the hills on the other, English, side. The location places the garden in its own, almost other, geography. 'It's kind of cosy,' Martha explains, 'but there's also an element of you that feels isolated.' The man who came here two generations ago 'would go off and do his thing' when he moved here, but his wife found it more difficult, Martha says. 'Apparently she would sit and stare at that hillside and it would just make her heart break. She felt trapped.'

That Martha ended up living in the cabin was by much

the same alchemy that she used to wind up with a job or a lover after a weekend at a festival. She was working at the garden as part of her training to teach at forest schools. 'I was what's known as the Kettle Bitch,' she explains, as we walk around an enormous twisted hazel tree. 'You have to tend to the fire and the kettle, basically.' Sometimes, Martha would spot a woman in the gardens. 'She was this busy bee, bright, blond woman, always carrying tools. She'd have secateurs in her holster and a chainsaw in her other hand, just casually slung over her shoulder,' she remembers. 'I was like, "This woman is amazing. I want to be friends with this person."' During lockdown, the holiday cabins in the garden were vacant; Martha and the woman were connected, and she moved in.

Martha ushers me into the cabin and it is gorgeous: a happy explosion of blankets and trinkets and jars of things on open shelves. An innately female nest. Her bed faces a picture window out to the gardens; there's another that lets you look down the valley from the shower. A wood burner sits in the middle. At a table, there is just room for two, the chairs padded with layers of warm things to wear. The space, she tells me, has been making her very happy. After seeing in the New Year with friends she self-isolated here for nearly three weeks. 'I got very used to this view,' she explains, pointing out of the window that floods the place with light, even on a grey day.

'Developed a bit of a crush on a squirrel that used to come sit nearby.' Being in the space so much by herself, she says, has forced her to make it home.

Martha has been working outdoors in different capacities over the past few years. She was a market gardener for a bit and a forest school teacher, a runaway from a corporate job in London that made her deeply unhappy, created anxiety that took years to shift and left her 'with a really intense fear of the natural world'. It took her time to discover a confidence in her surroundings, but also to 'uncover a comfort with feeling discomfort, that the natural world is so vast and inexplicable in so many ways that just being able to witness and experience it is amazing'. Because life immersed in and reliant upon the outdoors is not easy; there's a reason why humans have built themselves away from it. And yet we frame nature as a 'retreat', a means of escaping the problems seemingly constructed by our non-natural lives. We expect to be revitalised by spending time there without realising how far we've been removed from it, or how challenging that can be as a process. The weather can turn, the path can peter out, sometimes it is difficult to constantly bear witness and tune in to our surroundings. The notion that nature will endlessly astound us is as simplistic as the idea that it can heal us – for some people, that will be the case, but it's usually more complicated. Often we arrive at the edge of wildness expecting to feel some kind of awe, an immediate passage into a transcendent experience; when that doesn't happen, it can feel like a whole other failing.

Martha describes being flung into the isolation of the cabin as 'quite shocking, actually'. She had moved because she was missing the connection she thought would arrive with living in a place where, as she says, 'you have to so actively participate in life'. What she hadn't anticipated was that connection would be hard-won, and come from unlikely places. 'Some of it came from humans, but a lot of it came from everything else,' she says. 'I really formed a bond with the cabin that I live in, with the little mouse that comes in here sometimes.' Through that window, she has carved a deep relationship with the land, and how it has changed: the copper beech beyond it has just started to sprout leaves after months of bare branches, transforming her view. Such is the smallness of Martha's life here, and the contrast with the enormity of her surroundings, that she's found her thought processes deepen. She tells me about how she used to think about plants and horticulture before, and I recognise it completely: 'I used to have that urgency of "I must capture this thing and then put it in my brain so then if someone asks me about it, I can say, 'Yes' and give its Latin name."' Now, she tells me, she's keen to seek out different connections – ones beyond the mere collection of information, an accumulation of knowledge that may have to be deployed. 'I just like to go and experience them and then see how they change,' she says. 'I see it as a kind of, like a curiosity thing.'

Moving here, and into a more barter-based economy – she trades a day of labour a week for her accommodation – Martha has found the time and space to think about the outdoor world differently: not as something to know intellectually or in a way that could be used for profit, but a move towards a more meaningful life. 'There's part of me that has a growing anxiety about getting a bit older, and feeling like, "Oh my god, I've got no money, I've got no prospects" or whatever,' she admits. 'And then it's like, but here I'm living for free.' Martha says she toys with the sense that she *should* be 'saving loads of money', but then she realises that in being 'forced to slow down so much', she's learned to fully relax – something that she says she was never capable of doing before. 'This place,' she says, 'has profoundly changed me.'

What is the shape of her day here, I ask. Martha admits she has a lot of free time. 'So I'll wake up, I have my little tea. I'll make it and sit back in bed and look outside again. If I'm not working in the garden here, maybe I'll do some bits and bobs in my garden, or we've got some really lovely walks around here. It's nice to do that.' She's been painting, she says, and running; 'I dance around quite a lot.' Without the structure of a job or the company of another person, the recognisable elements of a daily routine slip out of place. 'I'm sometimes not very good at like, making sure I eat on time. I often feel like I wish I had a house husband who would do the cooking and tell me to eat and do the cleaning,' she jokes. 'That would be so great.' Again, I think of Abra, whose determination to renovate her cabin or sow her crops or chop her firewood often left her going without meals for days. Perhaps this is what it is to remove ourselves from the lives we've been told to live.

It has taken a retreat – both in the form of the cabin, and the removal of herself from a more typical urban existence – for Martha to learn new ways of living. While we are speaking, I recognise some of what she is saying: the desire to accumulate knowledge of plants, rather than allow myself the patience and time to inhabit and appreciate them; the challenge posed by societal expectations around money and stability to living in a simpler way. I admire Martha greatly for taking herself off in this way; it seems enormously brave to me. I also know that I can't, and wouldn't want to do it. I feel loneliness keenly enough living in a metropolis with my partner, and friends within easy reach.

•

Later, I meet someone else who makes me wonder how I would cope if I had been taken out of the city, away from my friends and my creative work as I knew it, for a year. Holly had answered my survey irresistibly – 'I was a drag king. Now I'm a market gardener!' – so I went to find her in East Sussex.

I'm driving down a one-road hamlet when I first see her, rangy and crop-haired. On this strangely windy summer's day Holly's wearing belted shorts and rolled-up sleeves, like someone from the past, captured in a photograph. The organic farm she's worked on over the past year is deserted; it's warm inside these vast, transparent rooms, and quiet out of the wind. Holly points out what she's planted, what she has propagated and sown. Before she came here to work – an unexpected departure from her master's studies, triggered by the pandemic – she had barely gardened before. Upon starting, it wasn't her horticultural inexperience that worried her so much as whether her body would handle the relentless hours of physical labour. She and her partner rented a cottage nearby and worked on the land. It's been an unexpected interruption in the academic, metropolitan life she thought she would be living. Rather than being drawn to the earth, and growing out of a need or a desire, Holly is here because it made sense at the time. 'You know, I had my path,' she says, as we sit on a bench outside the glasshouses. 'I do feel like I found this place by accident. But a good accident. I'll always be very, very grateful that this happened, because I think it's introduced me to something I'll definitely take with me.'

When Holly is in drag, she is Orlando, a persona inspired by Virginia Woolf's novel of the same name and whom Holly has performed as for more than four years. Orlando is fabulous and handsome and beautiful; he wears bodysuits and soft pink ruffles and eyeliner, and lip syncs to 1920s songs about gender-bending. 'Orlando was very, very important to me,' says Holly, when I ask her when he turned up. 'Because when I was younger, I'd been in an abusive relationship. When you're a teenager, you begin to learn that you're a distinct individual. But that process was made incredibly difficult by the circumstances I was in. So Orlando was a way of finding that person, I guess, and giving a voice to my queer identity and staking claim to who I was. I wouldn't say Orlando's a mask, because I think Holly and Orlando have always been very inter-twined. It's just Orlando is the most flamboyant I could ever get.'

I've found her on the cusp of change: her partner has taken a new job; together they will move to London, leaving the farm behind. 'You're seeing me treasuring every day,' she says, 'because I know there is an endpoint.' It's encouraging her to reflect. Holly never anticipated this year and soon it will be over. Moving to the countryside offered many things – big, arching rainbows above the glasshouses, new people, quieter evenings – but it also halted those that were big in Holly's life, namely academia and Orlando. 'Moving here, I couldn't be Orlando any more,' she explains, 'but I think that was good for me, actually. In many ways, to do drag is about being someone in the context of having an audience and putting on a show. Having a voice and being someone felt like an important endeavour for a long time, because it felt like I had to make up for having no sense of self for years.

'But now I'm four or five years older, and it was quite a good time to maybe leave that front behind and sort of find confidence in being just Holly,' she says. The sky has clouded over above us. 'You don't have to be anyone around plants, you can just be yourself. I think it was nice to step away from that and think, I'm good enough on my own, without the front of Orlando or without having something to say for myself or something to show for myself, just to be around plants. And that, I think, was important. It's a sense, a sense of quiet confidence, I'd say.' I've heard this a few times, while speaking to people about plants and gardens, whether formally researching or otherwise: that plants don't judge, and what is offered in exchange is a kind of freedom to be. People who find it difficult to be around others or struggle with negative thinking and anxiety, especially, relish the ease of gardens, how they enable them to be exactly as they are. I'd not thought about plants in that way; perhaps my own judgement on my ability to grow them shouted more loudly than anything else. But my body holds a certain privilege – white, cis, heterosexual and able – and it is easier for me to move around the world than it is for others but I am beginning to understand how gardens can be accommodating. For all the generosity of the soil – the fruit and flowers it throws up, the corms and tubers it can swallow – there is somehow always room for more human feelings. Children make dens from blankets and torches to feel safe. We can do the same things in our gardens: they are spaces that mean different things to those who create them, and those who observe from the outside.

Holly and I sit for a while. The wind has dropped and it's turned into one of those flat summer afternoons where time loses meaning. When we speak, there are times when she reminds me a little of myself a decade or so before; her colleagues on the farm are older, and she says they've urged her to stop worrying, that she has so much time to sort out what she needs to. 'To enjoy it,' she says. 'It's so important to hear. There are things I want to do, and I will do them, but it doesn't matter that it's not now and it's not maybe in the timeline that I had initially proposed. I'll get there because I want to.' Growing, Holly tells me, 'has very subtly but surely pervaded into lots of aspects of my thinking'.

Martha and Holly – both in their twenties – saw their experiences of relying on the land as life-changing things that neither had particularly planned for. It made me pine, a little, for that time in my own life. Had I spent so much time removed, so engaged with the land, I might have changed, too. The retreats they had found had not been easy, but in immersing themselves in isolation from people they'd gained a stronger connection with the earth and developed a fierce confidence in themselves. Spaces change – the cabin Martha lived in would eventually house somebody else – and meaning and sentiment can be as fleeting as an annual crop: we scatter what we want to see into the earth, and then find what we need to later, only to move on and leave that ground for another purpose. Retreat can also mean different things. I have always feared loneliness, and I hated feeling so isolated, but the ground – the garden – could hold it, if I chose to push that into the earth and let it germinate into something stronger. Retreat changed these women; they grew with it. If I could see this time as a necessary dormancy, then maybe I could too.

6

DULWICH

THE THINGS THAT I BURIED in the autumn are starting to emerge. The smooth noses of tulips appear above the ground; the dark purple *Iris reticulata* become the first flowers I'd planted in the garden to bloom. The daffodils I threw in sceptically in October – too much yellow in the garden, then, to contemplate much more – spear through the earth and I'm so excited to see some colour after the dirge of winter that I can feel it course through my body, bright and glittering. The sap rises and with it my anticipation. I start spending whole days out in the garden, keeping my hands busy to stop them from stiffening with cold. One Sunday, I finish up and turn to the east to see a full and creamy Snow Moon rising against the remnants of daylight. With a deep, unexpected satisfaction, I realise that the garden and I have somehow broken through; tentatively, the conversation we had been reaching for has begun.

In the garden, in life, I am teetering. Winter has been long and cruel; I miss the lights and the rush that usually comes with living in the city. I depend on the garden for it instead. In the dusky, too-early hours, I leave the gentle rhythm of Matt's sleeping breath and go outside: sometimes to the parks, where I have the whole place to myself and my bike; sometimes just to the garden. In both, I look for signs of change after a year when time has felt broken. New leaves, petals waiting in translucent buds, an undeniably different smell in the air. I spent so much of autumn pushing things beneath this soil. Now they have returned, they are everything and nothing as I expected.

I begged the past of every woman I was speaking to, but I hadn't spared a thought for my own. When time was listless I became fixated on the future: when we could touch our friends again, when we could board a train again, leave the country again. But sometimes during the conversations I was having, the young woman I had been would come to mind. If you plant a bulb the wrong way round, the shoot will grow beside it to find the light. Pull a bare-root plant from the soil and you'll see small leaves ready to emerge. The mulch I top the containers with was once plants. Even in these changed states, everything has been the same matter at one time or another.

I'd gone to Louise's garden to speak about her decision not to have children. But I began to see her garden – where she had tackled the buddleia, and held big plans for what might fill it instead – not as a child replacement but as an extension of herself in a different, more poignant, way. The conversation shifted slightly as Louise told me about why having a space of her own making – indoor, outdoor – was so crucial to her wellbeing. One of her previous relationships, she said, was 'horribly toxic, abusive [and] awful'. She described it, carefully, like inviting someone into your home – 'a sacred, sanctuary space' – only for the person to gradually take it over. Metaphorically, the person would move things or replace them, or take beloved objects until, 'eventually, it ended up being this situation where that person was basically just bulldozing the entire place. And at the end of it, I felt like I was sat there in a big pile of rubble.' The relationship left Louise in a mess – emotionally, mentally and physically – and one of the important steps in her recovery was renting a flat by herself, an eyrie-like place in a block in South London. She filled it with all her stuff, arranged it exactly as she wanted, and brought in house plants. 'It was just another layer of the house that represented things being well, and happy, and alive.'

Louise referred to that time of her life as 'ancient history', but it changed her thinking and what she values in life. 'You still think about it a lot. It doesn't hurt in the way that it did, but you still end up thinking about it a lot.' She and her partner have made a collaboration of their home, but she recognises the garden as part of the rebuilding of self 'brick by brick' that can come with making a safe space of where you live. 'It's something I find great solace in, whenever I'm stressed, whenever I'm upset, whenever I'm kind of lost,' she said, 'I can come out here and get lost in it.' Later, Louise shared her plans for a greenhouse tucked in the corner: for all the work she and her partner had put into their home, it was still something built together. A small glass building could be something she alone would occupy, she alone would grow in. A haven within a haven, a place of her own creation. She explained in five words: 'I just want that space' – and I knew, instantly, how much more a structure could be than plants and square footage.

I was unable to stop thinking about how Louise had built herself a new home – and with it a stronger sense of self – after suffering from an abusive relationship. I was furious at the injustice of what she'd been subjected to, I was sad for the shadow self she'd had to live as and I was moved by her recovery. The flat she moved to crested the next hill along from where I had lived in Treehouse; while unsure of the timeline, I thought of us as strange twins. Two women finding their feet, defining their own spaces, after the supposed plan changed course. Two women, now, learning to share again, to build homes and selves again, with partners after knowing the value of one's own room.

•

Over the winter I thought about being in the room I'd rented while I was younger. Specifically, I thought about being there one afternoon, after a nap, and waking up to snow. I remember it as a magic trick, the kind of event that brings out smiles in nineties children raised in southern England. Snow was rare and beautiful. Fat flakes, deep blue skies, orange haze from the streetlamp. At the end of a week where snow had fallen in bitter, persistent showers, I woke before dawn with a determination to read my diaries from that moment, that afternoon snowfall. I had a clear picture of what I'd looked like at that age, but I couldn't grasp what I'd felt like or how I'd thought about things.

At the time, I was in a relationship that I would later understand to be emotionally manipulative, possibly abusive. It was an amorphous, quietly devastating thing that, over the months we were together, systematically diminished my self-esteem. I often felt I lived two lives: one with the man, and another, happier one with my friends and our parties and our dancefloors. From the off, I was attracted to him because he seemed unimpressed by nearly everything but me and because I thought he was very beautiful.

I made myths about him and kept them to myself. Even now, I think back on that relationship and understand much of it to be the kind of troubled love story that falls into films and novels, rich with a weird, shimmering romance that seems out of kilter with the everyday comfort of the good, honest love I have been fortunate to know. It took years for me to acknowledge that these things had happened.

The diaries weren't hard to find. My past is packaged into a handful of reused cardboard boxes, which are dominoed into the bottom of a sideboard. I slid one out and a handwritten label confirmed my suspicions: *'photos, notebooks etc.'* I took a knife to the thick layers of parcel tape and they were sitting at the top: three fat Moleskine notebooks, covers dimly shining like eggshells. To open the first was to find the address of that house, with a bloom of inky water damage on the page corners. But what followed was less transportive: pages of email addresses, notes for juvenile newspaper features.

This was how the evidence of that afternoon appeared: a small paragraph in handwriting tighter and neater than it is now, folded between lists of ingredients and prices at the shop I worked in. And I had written about what I'd been thinking of recently, brought on by snowfall. Time had warped the details – it wasn't an afternoon nap, but a morning one; I suspect I'd snuck back to my house early to avoid explaining the situation to my flatmates. 'The next morning, tiny dry snowflakes blustered into my face and I fell asleep. When I woke up, it was entirely white outside.'

That was the extent of it. The rest of the notebook unfolded in the everyday: to-do lists, notes on Romantic poets or John Donne and, for a good chunk, pencil-dashed details of a few weeks spent rattling around Eastern Europe with a friend. Occasionally he left notes in the pages, but nothing that seemed untoward out of context. Whatever I'd hoped to glean from my younger self wouldn't be found here. I closed the notebook, put it back in the box but managed to linger in the present.

In the heaviness that followed, I wondered why I'd gone hunting for evidence. I've never been much of a diary-writer; the notebooks I keep now, like the ones I kept then, are mostly full of orders and lists, a means of laying out thoughts and ideas rather than memories. As the sky brightened, Matt nudged the study door open and asked if I'd like tea, and I felt ashamed to be caught bundling boxes back into the sideboard, guiltily raking over things that took place firmly in the past.

In the years since that relationship I've thought a lot about what it is to be believed, and I've thought about the shapes and the spaces we are allowed to occupy as women. The lessons society teaches girls suggest rape, assault and abuse happen in small, neat boxes of circumstance: strangers, dark alleys, short skirts, too much to drink. The reality is messier than that, womanhood is messier than that. I think, now, that I carried on because I packaged these events up as I did those notebooks. Put them inside boxes of denial and coping, where they sat for several years until, gradually, I pulled them out again and had another look. I spent my adolescence feeling gawky and undesirable and when it happened I remember telling myself that this was what male desire looked like. Pride and shame – about the entire relationship, really – ushered me into a silence that I kept out of habit.

It would be trite and untrue to say I gardened alongside these revelations, but they did unfurl during the first years that I started to grow things. I think of that time as one where growing up happened to me. I began a new relationship and we bought a flat together. I moved into it, we moved apart without me realising. It was on that flat's balcony that I grew a cocoon around myself and questioned the person I was. It collided with a growing awareness of my feminism, of glass ceilings, everyday sexism and gender injustice in the industries I wrote about. I suppose I woke up. Not snowflakes this time, but the grey slush left behind.

Hearing Louise talk about space made me re-evaluate those I had made and those we make as women in defiance of what is expected from us. Girls are raised with a keen and unavoidable understanding of beauty standards, but when women make gardens, we make spaces that we define as beautiful on our own terms. I thought of Louise's flat filling up with houseplants, rebuilding her self-esteem with every unfurling leaf. I thought of the herbs I bought from Columbia Road market and prodded into growth in the side-return of the house I rented a room in during my early twenties. I thought of the tomatoes and nasturtiums I let rampage all over that first balcony, even as I signed the papers to sell my share to the man who left me. I thought about the women who crossed boundaries into neglected land, defied the rules in taking care of it and transformed it into community gardens with courage and creativity. These plants were beautiful, but they were even more so because they grew on their own terms and they took up space that nobody told us we could inhabit. I didn't garden to recover, specifically, but when I admired a concrete space smothered in the leaves of seeds I had scattered, I felt a strength and a resilience that nobody – no man, no abuser – could take away. To green a space is to occupy it. To spring life from dirty grey concrete is to stage a takeover. To garden is to cultivate a superpower.

•

In recent years, people from my past would reintroduce themselves to me as gardeners. I'd open an inbox to find queries about courgettes or be shown proud photos of wildflower plots. A decade ago, we had been hungry for the city, for its ladders and industry. Now we just wanted to make a place to get away from it. I'd been aware of Rhiannon's burgeoning North London garden for a while; she would post questions about it on Twitter from time to time. We'd met several years ago, when she and a friend were graduates. For a short time, we'd drifted into one another's circles, but I'd not seen her beyond social media or email for years. In the decade that had passed, we'd become women.

Rhiannon's rented the same flat for all that time, a small fact that's a rarity in London, a near-impossibility for people our age. And it's here that I meet her. She opens the door laden with clinking bags: the flat is upstairs, the garden is down a path; we are having a kind of picnic.

The garden was a mess for the first five years that Rhiannon lived here: an intimidating nest of brambles that would get cleared every now and then by an enthu-siastic visitor, and then grow back. There was wine to drink, parties to attend, careers to build. The flat was rented – she didn't see the point in investing time nor scarce money in it. But at the end of 2015 Rhiannon and her husband were caught up in the Bataclan terrorist attack in Paris while on a press trip. They were outside on the street after dinner when her mum called with the news. 'I remember my knees just buckled,' she says. After searching to find somewhere that would take them in, the pair were stuck in a bar 'for hours and hours, not knowing if they were going to come and shoot us or not. We spent the whole time just looking at the door.'

Rhiannon tells me she thinks she 'would have been okay, psychologically', had she not already been managing PTSD from an assault that happened while she was a

101

student. These days, she's almost blasé about the impact of 'a random man trying to strangle me'. Had it not happened, she wouldn't have moved out of the area, befriended another young woman with whom she wrote a book, found the flat that contained her future husband. Still, she says, with a kind of curiosity: 'my twenties were kind of bookended by two traumatic events'.

The result was that Rhiannon developed agoraphobia – an extreme fear of leaving one's own home – rendering her unable to get public transport or visit friends in bars or restaurants. 'When I did I'd be in a state, I couldn't sit near the window. I was constantly just expecting to get shot, basically,' she explains. It was a visiting friend who pointed out the overgrown plot outside – what a waste it was not to tend to it. 'It was my husband who said, "Well, it's not like you can leave the house, so we might as well make you somewhere you can go."'

Together, they dug up every bramble root by hand, unearthing roots the size of a human head; a scrap of paper was drawn up by a colleague with design experience: a little wildflower patch, somewhere to sit, pots around the side. We sit among it, and I imagine the roses and the lilac that will bloom in the coming weeks as Rhiannon describes them. She didn't have much cash, but she had ambition: rose cuttings from Poundland, a small magnolia 'that will take about 50 years to grow'. Rhiannon's birthday is in June, and she planted a June garden – one that swells with midsummer. She found new, horticulturally-minded conversations with her grandmother and uncle; they sent her cuttings through the post: lavender, geraniums and penstemon. Offerings from around the country landing on a London doorstep. She learned that her father loves roses.

Making the garden helped Rhiannon be less afraid. 'I was so frightened of everything,' she says. 'I mean, everything. A plane would fly over and I'd think it was going to fall from the sky; my fight-or-flight was going off constantly.' The ground gave her something physical to touch, something else to 'obsess over'. She'd start thinking about what she was going to plant, where she was going to put it; her curiosity encouraged her to explore the local neighbourhood, see what was growing there that might work in her own depleted soil. Rhiannon had no interest in gardening when her husband cleared the plot, despite it being a family passion, but as she coaxed herself down to the garden she found she knew the names of things, had strong, instinctive opinions on what she liked and what she didn't. When she went to ground, Rhiannon uncovered parts of her life she'd forgotten, a heritage she never knew she had.

It's strange to hear Rhiannon speak about such fear in this space. Even now, months away from midsummer, she conjures the garden it will become with gentleness. She's very much a suck-it-and-see gardener: bargain dahlia bulbs chucked in pots, love-in-a-mist scattered around every fortnight like her grandmother used to. I can tell it is a garden that tumbles, that holds long dinners under heady lilac, laughter puncturing the not-quite-darkness. She has deliberately encouraged the roses, clematis and honey-suckle to co-mingle. This plot, grown from fear, has been one free of meticulous control. Rhiannon's brother is severely autistic, and she learned to grow up – to become the writer she is today – in a chaotic environment. When she writes, she tells me, there's always a tension: no matter her perfectionism, the words will never be what she envis-aged. But the garden is free of that, Rhiannon says: 'I can't get preoccupied by stuff that's not working out how I'd want it to. It's my space for just being like, "Go on, do your thing. Let's see what happens."'

She says she loves planting stuff and forgetting about it, and I think about how the ground can hold so much more than the matter we put in it: fear, future, the family that made us. For five years the patch of brambles we are sitting on was forgotten – perhaps, if she and her husband move, their work here will be too. But then there will be another garden, another escape, another cradle for recovery.

•

I cycle over to Megan's place one grey afternoon, and notice the seedlings perched on the front-room windowsill as I ring the doorbell. She invites me to push the bike through into the hallway, dark unruly curls tumbling around her impish face, and tugs on an anorak before taking me through the kitchen to the garden. Occasionally, the clouds spit at us, and we spend the next hour or so pulling up our hoods and pushing them back again.

Like Louise, Megan had responded to my survey. Like a lot of people, she'd turned to the outdoors with the onset of the pandemic: an actor and director in her twenties, she had been on the cusp of her first big break. Having closed a play that was due to open ahead of the first lockdown, she retreated to her mother's garden in Oxford. There, in the unending sunshine of that spring, Megan came to see the garden as a space of freedom, a space away from the turmoil unfolding beyond its boundaries. After existing in a competitive career in a competitive city, Megan relished the ambiguity the garden offered – there was 'no right or wrong' there, she told me – and the 'particular kind of tiredness' she felt in her body after hours of physical work. When she returned to London in the summer, she did so with armfuls of plants and a determination to make her own space, as her mother had. Gardening opened up a new line of communication between them.

We sit in the garden Megan has made from an overgrown plot: there's an apple tree, just out of bloom, and a cold frame full of growing things. Nasturtiums leaf in a trough by my arm. Most of what she's growing is in pots, rather than the beds, so she can take her garden with her if she moves.

I ask what Megan's growing on the front-room windowsill and she pauses. 'I'm going to be quite open here,' she says. Earlier in the year, she started seeing a counsellor. Her flatmate had found the service, which was affordable, and both embarked upon some sessions during the winter lockdown. 'I was sexually assaulted six years ago,' she told me. 'And I kind of didn't realise it or talk about it until then, when I started to talk about it.' After every counselling session, Megan really wanted to be in the garden. 'There's something about getting out of my head but also letting the thoughts sit,' she explained. 'When something to do with it came up again, I spent the whole day in the garden. I was like, "I'm just going to find things to do." I wanted to plant seeds, it was going to be my marking of this thing, and me changing because of it.' She liked that she was playing a part in the growth cycle but wasn't in control of it. That the seeds would germinate, grow, produce more seeds and die, leaving new, different flowers in their wake allowed Megan to see what she had been through differently. 'It made me feel that part of me wasn't unuseful; it wasn't necessarily what I wanted, but it wasn't a bad thing,' she told me. 'It didn't have to be something that never grew again, a part of me that didn't grow.'

When I listened back to the recording of our conversation, I heard my voice crack as I thanked Megan for sharing with me. I told her that I had lived that, too; that I hadn't spoken about it much, and that I was still figuring out what those events meant in my life. We were kind and awkward about the chasms of vulnerability that we'd opened between us, in the back garden of a scruffy terraced house in South London. Our words tripped over each other's only to sit in silence. I was so grateful for it, this unlikely space of acceptance. Megan told me that for her, her experiences and her relationship with the garden were firmly connected: the ground gave her something to hold, to indulge in and escape to. I was still not sure if that is what it was for me, but I felt so validated by her insight. For years, I had swaddled my survival under shame and silence. Increasingly, I wanted to let some air in, let it be something that grew into life. I had been looking towards women who had experienced what I thought my life might yet become – motherhood, middle-age and beyond – without realising that there was much to learn, and even forgive, about the womanhood I'd already lived.

The fingerprints of abuse still lie on my body. Small and featherlight, banal and unique. A scar that left me marked and turned me from an innocent, of sorts, into a survivor. While I had no choice over their existence – they were imprinted upon me – I could choose what to do with them. Whether I would let them define me, whether

I would keep them secret, at what point I would put them on show, if at all. I had existed for a good while without being aware of them: our brains are good at burying trauma; my mind was adept at denial and reconfigured the facts. But what I think those fingerprints did do was to galvanise me. The muscles beneath the skin they touched grew hard and sinewy. They became strong and fierce and resilient and encouraged my voice to follow suit. I stood louder and taller for women's rights, called out sexism and gave it more scrutiny. Where once my experiences – that victimhood – made me lonely, I learned to recognise them as a ticket to a sisterhood. That I was one of millions of women who knew what it was to have their choice taken away, that I could recognise their pain and turn it into power.

7

ARTEMIS HOUSE

I'D HELD ON TO THE notion of breaking ground since the beginning of all this, of gardening as rebellion and of making change. Ground can be broken from above, with the blunt edge of a spade or the swinging thud of a mattock, but I tended to picture it from below: the gentle, determined force that pushed the curved neck of a seedling stem through the soil above. It was a power that always impressed and surprised me, that emergence of life and potential from the earth.

Women, I knew, broke ground when they gardened. Historically, gardening was yet another activity that – for middle- and upper-class women, at least – wasn't deemed acceptable until the 19th century, even though women of lesser means had been growing to feed and care for other people for millennia. Although more women than men study garden design, the majority of RHS Chelsea Flower Show gardens are still designed by men, who take home more of the medals; there has never been a female lead presenter of *Gardeners' World*. So it continues.

But women were making change the way they always had: quietly, feverishly, at the grassroots level, with babies on hips and caring responsibilities. I wanted to speak to women who had turned to their gardens and the green spaces around them and seen them as vehicles for change. Since I'd gained access to the garden, I'd wanted to do right by it. I'd set up a compost bin before I got to the ground; I'd ruled out using peat in the garden and picked up what I could second-hand. But as the months shifted, and the garden grew, I realised that being a good custodian of the earth was often more in what we didn't do – what we left alone to flourish and take up space – than in how we intervened.

•

The end of May sees heat creeping in, muggy skies and surprise sunshine. After weeks of relentless rain and bruising winds the garden is able to stand still. I remove the cloches from the sweet peas, usher the seedlings back outside, bring the pots back onto the table. The house takes a while to warm up, and I leave it and find South London ripening – crop tops and loudspeakers and the excitement that precedes the hedonism of new summer. I garden with a mojito in hand, edging the lawn before a dozen friends come and sit around the table the next day, talking and laughing over one another, until we huddle under blankets and decide it is time to go inside. While nobody is watching, the tight, hard buds of the peonies soften and swell. A sunny day later and the first one is out: a near-offensive bright pink. They open and close with the day, and they make me think of the previous summer, when I ordered the bare-roots within hours of exchanging contracts.

The warmth spreads into early June, making me sleep later and stir sluggishly. Sometimes, I wake in the night and think of the garden or have dreams about its progress, rising and going out there in plastic shoes and a dressing gown to peer and prod and uproot and tie in and marvel. At that time of the day, it's just me and the birds. As the hours yawn on other noises join in – the man next door running his rhymes, the bloke down the road with the sound system, the neighbours who sit in their windowless outhouse and talk and talk and talk. But before 7 a.m. I alone get to watch the new sun hit the shadier corners of the garden; I open the back door and I am blinded by it, keep walking until my eyes adjust and I see what has happened overnight. Sunflower seedlings, calendula seed-lings, dill seedlings. A new orange poppy from one plant, a first white one from another, wearing its fuzzy casing like a pillbox hat. A first flower on the hardy geranium I planted bare-root. A flush of aphids being patrolled by the ants who farm them on the *Ammi majus*. Drunken alliums; perhaps they'll perk up during the day. I tie in the lumbering sweet peas, pinch off their buds – I want them taller, still.

As the day stretches I repeat this walkaround, sometimes crouching low to be dwarfed by the tall leafy growth. Days earlier, a friend commented that the garden was 'all very *green* at the moment'; they asked whether that was what I was 'aiming for'. Now the colour is appearing in a gaudy rainbow: bright pinks, searing oranges, unexpected violet. All that intention to maintain a colour palette and here I am with a whole paintbox blooming before me. Growth, growth, it fills the month. Things have turned up and, like friends of friends at a party in full swing, I do not recognise them.

Beneath it all, a lawn that grows long and lazy. Buttercups, some pink thing, drowsy seedheads. I love the tuft of it, soft beneath my feet. I can think of nothing I'd like to do less than mow. The yarrow fills the herb bed with yellow; the blue-black of the salvia in the middle. After a long and peaceful respite from slugs something is back and munching. The thinnest of spider webs has been spun between the sweet pea stems that have outgrown their stakes. A glistening green beetle has clocked on to my poppies. It is good, is it fine; this garden isn't just mine to take in. I watch them all arrive.

The energy behind this does something to me. A

swelling, a softening, a letting go. I can feel it stirring. I wonder if this is the same force that galvanises women to do things nobody expects of them, to grow and create from overlooked land, to make change.

Ground rarely breaks neatly. We disrupt the complex structures of the soil, we trample the things that are growing. Those who want to make change believe the mess is worth it for a new way of doing things. I'd not sought to make a change when I gave in to my curiosity over why women were drawn to the land, but I could feel it happening – inside my mind, outside in the garden. As squirrels and slugs appeared to eat the plants that hadn't been there before, so doubts crept into my mind; often, I was so preoccupied with reaching a vague destination that I missed that much of what I was learning happened in the process.

Those women I saw or read about who were going to ground to make change seemed deeply courageous, in a way I didn't think I'd ever feel. They embarked on something without knowing what the result may be, holding on to it in the face of doubt. Sometimes, things would go awry or a project would end, but the shifted parameters were evidence of the bravery that had started it all. All of us have to live in a world built upon deeply ingrained patriarchal, capitalist and white supremacist structures; even to imagine a way of living that counters these, let alone undergo a process of bringing it to life, is an admirable thing.

Many have tried. Land has been a potent means of power for women for decades: in Texas in 1879, women achieved a financial independence years before it was the norm through the foundation of the Woman's Commonwealth, ostensibly a Bible study group that used the land women inherited when their husbands died or abandoned them. It became a feminist sanctuary of sorts: a formerly enslaved woman was among the 50 on record; others had escaped from abusive husbands. They ran hotels and farms and founded a library. While the community declined when its founder, Martha McWhirter, died in 1904, it lived on with its last member until 1983. There are land-based women's communities in New Zealand and Japan, where the members live on buses. Greenham Common Women's Peace Camp existed for 19 years, despite sustained local aggression. The Carpathian Mountains of Ukraine are home to the Asgarda, a women-and-girls-only martial arts training school that has educated more than 1,000 pupils.

The Women's Land movement that bloomed across America in the seventies shows how different ways of living and the attempt to create a utopia removed from the constraints of conventional society can unravel. Emerging from the lesbian separatist movement that bubbled up in the late sixties and a broader shift towards land-based living, these intentional communities took different shapes and structures but were nevertheless a movement. There were enough to support the existence of *Country Women*, a magazine typewritten and hand-illustrated by Carmen Goodyear, a filmmaker who has lived in an all-female community in Mendocino, California, since 1968. *Country Women* gave women practical advice on installing solar power and caring for animals. Vestiges of the communities remain, but five decades later the politics that fuelled their creation have become thorny. At a time when discussions about gender identity and trans rights are so loud they become newspaper catnip, and the urgent plea for trans rights is taking to the streets, those remaining communities that define themselves as women-only are having to confront their history. As Keridwen N. Luis, the author of *Herlands: Exploring the Women's Land Movement in the United States*, explains, she was prevented from spending a year living in such a community 'by an issue that stops many women – lack of funds or means to support myself in a highly rural area'. Women's reproductive rights have never felt more threatened in my lifetime, and the crunch of late capitalism feels increasingly exhausting, but the women who escaped similar strictures in the seventies are now in their seventies – and we can't afford to follow in their footsteps.

I'm also not sure I'd be able to. The projects I encountered in search of finding women who turned to the land to make change often showed me my own prejudice. I found myself walking away from a school that used the edges of a woodland as its only boundaries, woodsmoke in my hair and on the collar of my coat, deflated that I'd struggled to engage with the work being done to offer a group of children a different way of being educated. I was frustrated; I felt I'd failed, that I'd come all this way to glean so little.

Afterwards, I drove west for a few hours and headed for the coast. At that time of year the hedgerows were lush and filled with darting birds. The deep pink of valerian and arching lilac spires of foxgloves buffeted against the air from the traffic. I thought about how much of my resistance to what I'd seen was due to my own resistance to change. I am the daughter of a schoolteacher: a neurotypical, academic-leaning people-pleaser who actively enjoyed school and grew into an adult soothed by order and routine. Being there reminded me of some of the listless, freewheeling summer schools I was occasionally put into as a child, which left me keening for someone to give me something to do to fill the time. But the children I'd seen loved the fact they were there.

Constant change, the school's founder told me, was a 'core principle' of the project. 'And learning how to work with it and be in relationship with it.' In that light, it made sense. Indeed, any initiative that tries to uphold a more flexible and generous relationship with change makes sense. As our systems – of capitalism, of white supremacy, of patriarchy – are increasingly shown to be connected in contributing to the climate crisis, embracing change and fostering a relationship with our natural systems felt as good a solution as any.

I was disappointed with my time at the project because it hadn't met my expectations of what I felt it should have been. I wanted to see a woman participating in a radical solution, born of the earth. Instead, what I'd been shown was that change doesn't happen quickly or beautifully and that the attempt to make something better was not always going to be elegant. I realised that maybe I didn't understand it because I'd only known the conventions of a society that needed to shift, and it was going to take time and thought – and the generosity of an invitation to bear witness – to see it as change as well as just disruption. We don't always see what our actions leave behind.

Carmen Goodyear is in her mid-seventies now, and still lives on – and from – the same farm she moved to from San Francisco with her wife, Laurie York. The communes have drifted away, but, York told *Vogue* in 2017, the legacy remains: 'Women taught each other, empowered each other. It's a dandelion effect. That early consciousness-raising was a time of planting seeds. They blew into the wind and spread everywhere.'

When we go to the ground, it can teach us different ways to live. The more time I spent with women who had turned to outside spaces for meaning and connection – to Martha, in her cabin; to Maya, who studied herbs – the more I realised how many of the structures I'd taken for granted had been recent constructs, built and designed by men to elevate us away from the relationships we'd nurtured with the ground for centuries. In Tudor times, 'old wives' were respected for their knowledge of plants; within a century they were dismissed by men who had written down their knowledge. Now we call gossip or rumour 'old wives' tales'; of course we've lost their stories.

•

After centuries of scientific medical development, herbal remedies linger in the fringes of healing. Anne is a herb-alist and a gardener, among many other things, and a pioneer in the field of alternative medicine. She's approaching 70, and has been studying, growing and working with herbs for more than 40 years. Along the way, she tells me, she's encountered repeating patterns of interest and scepticism in her work. She seems to have long since risen above caring what anybody else thinks.

When I first heard of Anne's clinic through a friend I imagined it to be somewhere imposing and gothic – with, for reasons I can't explain, a turret. Instead, Artemis House is a hill-hugger of a place, cleaved from pale Cotswold stone in the heart of a swooningly pretty village. The birdsong is the loudest thing. Inside, shoes come off at the door and there is a sanctity in the air as I walk down a wooden, open staircase to the garden room where she practises. I'm conscious that I'm occupying a unique position: I'm not a patient and I'm not seeking healing so much as understanding, although the two must come to overlap here. I've come here to learn something else from Anne – about her life, her practice and what plants mean to her. She looks – and seems – far younger than she is. Her auburn hair softly frames her round face; she peers over a pair of glasses at me with brown eyes, quizzical. At first, Anne's a little prickly, unsure of my dictaphone and the questions I'll ask of her. I'm to explain myself before I expect her to do the same.

Once I have, a remarkable story unfolds. With stillness and a growing warmth, Anne starts at the beginning of her life – a childhood in which she was sent outside to fetch herbs for her mother, who would cook with them; an adolescent fascination with Indian philosophy aided by tuition from a Buddhist monk. Anne was expected to go to university, but she put off her studies by travelling overland to India at 18 and staying away for a year. Comparative religion at Leeds struggled to compare with backpacking. 'You know, I'd been hanging out in temples only to come back and read these books that moths would fly out of, they were so dry and boring,' she tells me. 'People thought I was weird at university because I meditated every morning.' And so she didn't last long at university. When she told her tutor she was dropping out – 'this is way too boring, and I've got things to do' – he replied that he envied her, he'd like to run away himself. A trip to South America got sidelined after Anne read John Seymour – the author of the bestselling *The Complete Book of Self-Sufficiency*. 'I had a very strong sense that if we lived closer to the earth, then we would be more in harmony within ourselves,' she says. A 'sweet little wooden' house came up for rent on an island off the coast of Essex. 'I just said, "Right, I'm going to take it." So the money that I'd saved up to go to South America, I just sort of invested in my way of life.' Anne was one of five people who lived on the island, which was only accessible at low tide, that shifted every day, so steady employment was impossible. At 22, she immersed herself in a necessary education of survival: 'I just set to digging up the lawn in my garden. I had my *Food for Free* book and went out looking for things that could augment what I was growing. I wasn't going to be self-sufficient, but I was going to try and understand what was growing wild around me.' There was sea spinach and rock samphire and fish and cockles. 'And just me, living on my own in this little cottage. It was great.'

As Anne learned about plants – through foraging, through growing, through reading and eating – she started to think about balance: if we could be balanced physically, surely we could be balanced emotionally, mentally and spiritually, too. She learned that plants had medicinal benefits beyond their edible ones. 'I had a sense of the bounty of the earth, and that everything we need to heal us on every level of our being was there – we just needed to learn about it.' In time, she did: the money eventually ran out on the island, and Anne took a job at a herb farm. 'I finally ended up going to South America with a friend,' Anne says. 'And I think somehow, sitting on the top of the mountains in the Andes, and just having time to reflect, it seemed like maybe that was the thing that I really wanted to do with my life: to go back and study herbal medicine. I spent some time meeting *curanderos* and witch doctors and so on in South America. And I had the opportunity to stay and study and apprentice with a *curandero* in the jungle. And I just thought, actually, I'm from England, I need to go and do it my way.' Four years of herbal medicine training and 40 of practice have made Anne a pioneer in her field. She still, she says, doesn't know if she made the right decision.

Anne says herbs are her 'medium'. They offer one part of a complex and multi-disciplinary healing practice. But she puts her love of plants down to being 'a beauty addict. Really, really. Not that my house is beautiful at all, but I just think plants are beautiful. Things like roses.' She points to a posy sitting in a glass on her desk. 'Look at this perfect flower arrangement. Somebody brought me this today. It's beautifully artistic, isn't it? I'm drawn to that kind of beauty. I don't necessarily go to art galleries, I'm not a painter. But my thing is herbs. I can't imagine being a herbalist without a garden.'

Anne has always created good gardens, perhaps starting with that one on the island. We speak facing the one that unfolds behind Anne's back, through the sliding glass doors, across the patio and beyond. Once she was married – to another herbalist – and living in the Cotswolds, she made a herb garden so splendid that it 'was on television and *Gardeners' World* and *Country Living* and goodness knows where'. She left it, and her husband, 20 years ago and moved to the house we're sitting in now with her three daughters. 'That's why I call it Artemis House, because Artemis is the moon goddess. And she's the protector of women. So this is Artemis's garden.'

I have spent evenings in blush-pink, women-only members' clubs; I spent the later years of my adolescence assisting women in the changing rooms of a smart boutique; I have made instant, loving and short-lived friendships through swapping lipsticks with drunk girls in nightclub toilets. But I have never been anywhere as instinctively female as Anne's garden. I can feel it in my bones, this unerring sense of womanhood. The smells of it – of the herbs, of the flowers, of the damp and the bloom – and the potential promise of wildness are ripe in the air. 'When we came here, we were four women,' Anne explains. 'My youngest daughter was seven, and I felt like it needed to be a woman's garden. Artemis is fierce. I didn't realise that, but she's pretty fierce. It takes quite a bold man to walk through the front door of this place.' It's a place, she goes on, that is run by women – her assistant Jayne, her three daughters who still help out and stay for stints; Katie, who helps hugely in the garden and Anna, who runs Anne's courses. 'This is a powerhouse of women,' she concludes. One that is so strong, it has no need to nail any colours to a mast: you can feel it surrounding the place.

We stand in a tunnel of foliage. It's not much larger than me: the small, frothy flowers of a pale pink rambling rose graze my scalp, adventurous vines push against my bare arms in their search for light. My knees nudge against purple allium heads, swaying at a tilt, my feet step over herb robert furiously in flower. Raspberry leaves, alert and sharp, whisper at my skin. It's dark in here, the dull white of an overcast midsummer day swallowed by the season's growth, but it smells unlike anything I've ever encountered: close and heady and strong. Roses, but something deeper and earthier too. Something animalistic and fertile, something unashamedly pungent. Anne gently pulls a few small leaves from a low-growing plant, rubs them between her fingertips and offers her hand to me. 'Chamomile,' she explains. I tell her how overwhelmed my senses are here, and she laughs: 'It's a birth canal!'

Only half of the roses are out at the moment, Anne explains, as if my experience wasn't intense enough. And

123

they're there because they are helpful in childbirth. 'They're amphoteric, so if your uterine muscles are over-relaxed they contract them, and if they're over-contracted they relax them,' she says. 'And yet they're a symbol of the exquisite delight and pleasure, the amazing perfection of giving birth to a child. But it hurts,' Anne sums up, with a knowing laugh, 'so there are thorns.'

'When I came here, there was nothing, just lawn sloping down and a path through the middle,' says Anne, as we leave the tunnel to look across to the rise of the hills beyond. 'At the bottom there was a swimming pool. It meant I had a blank canvas.' The pool was covered and the area of ground around it became the foundations of the 'village' that sits there now – a cluster of wooden huts and cabins where Anne teaches and students can stay. On the lawn, a large spiral was drawn out and measured exactly, hunks of grass were removed and Anne laid out her different gardens along it, each a stage on 'a journey through a woman's life expressed through herbs'.

Anne points out the herbs as we walk around the spiral – there's climbing jasmine to represent entangled lovers; borage 'for courage, which you'll need to get married'; soft pink marshmallow for breastfeeding; 'dark things – red sage and skullcap' for the menopause; and 'foxgloves, that any self-respecting crone should have in her garden'. The enormous, umbrella-like flowers of angelica tower above us, representing the importance of 'supporting ourselves in old age'. Hyssop and lavender are there to help with forgiveness at the end of life – of what we've done, of what we've not, of ourselves.

I have never been anywhere quite like it, where the language, meaning and purpose of plants is so woven into the ground. It's the kind of garden built to resonate with people – especially women – whatever their stage in life, whatever they've known. In the marriage garden, I tell Anne that I will be getting married in the next year and she says you're supposed to plant myrtle outside your house when you do, but that it is slow-growing. Standing here, childbearing and breastfeeding exist in a strangely tangible hinterland; the menopause seems impossibly distant. But Anne has built this garden from the middle of her life, looking both forward and back. The garden, she tells me, 'is the place that I feel happiest. I like that feeling of working on the earth even though I'll be 70 in a couple of years. So there is a sense that I need to keep doing this to keep myself strong.'

At times, Anne reflects on her four decades of practice; she's teaching a lot at the moment – 'because maybe at some point it might be time to stop. I don't know, but I should probably pass on as much as I can that I've learned.' The same curiosity – that same nascent power – that fuelled a year in India, when she was a teenager, still burns now. 'I still feel that there's so much that I don't know,' she tells me. 'I'm not going to jump on a bus and go overland to India. But I'd rather sort of quite like to adventure into the widescreen stage of life in my garden.'

•

I arrive in the foyer of Hackney CVS (Council for Voluntary Service, although nobody calls it that) – a squat, slightly depressing little 1960s building on Dalston Lane – wet through. A thunderous shower belted down during the 10-minute walk between here and the station. Along the way, I've huddled from the weather in doorways of the fashionable cafes and bars that litter Dalston's main drag and made myself late. When I reach the reception desk, I can't see Clair, and worry that she has gone home. Clair runs Rainbow Grow, an LGBTQI+ community growing group that operates from the courtyard beyond the foyer's doors.

She clocks me, and declares – astutely – 'Well, you're not going to get any more wet.' A tour of the community garden unfolds. Clair's difficult to pin an age on, but I suspect she's in her mid-sixties. I follow behind, meekly, as she points out thyme and brassicas and wheel-chair-friendly planters at a rapid lick. Rainbow Grow is at the start of this year's growing season, but you can see the crops to come. True to its name, it has been laid out in colour order – there's the dark purple of kale, while orange is offered by nasturtium – crops and flowers happily co-mingle here. This is a space for members of the

LGBTQI+ community and their allies to find one another and an opportunity to garden. 'A lot of young people who didn't have their own space, three of us who are retired, a couple of people who have flexible hours, students,' Clair rattles off the group's membership. 'We've had people who are seeking asylum, we've had people from a lot of different cultures. It's intergenerational, and a different way of socialising that's not at a pub or a club. It wasn't around drinking.' This is both a welcoming space and a fairly tight ship: Clair points out a trough brimming with enthusiastic seedling leaves, lightly tutting at the new volunteer who was overzealous with sowing lettuce seeds.

Clair lives further along Dalston Lane, and has done so for nearly 40 years. It's social housing, one of those well-proportioned Georgian terraced houses that now shift hands between private owners for more than a million pounds. It has stopped raining, but the deluge has left swampy puddles behind. Clair shrieks at these and whips out her phone to video the cars passing and spraying filthy water onto the pavement: more evidence to add to the case she is bringing, with fellow residents, against the council regarding new traffic measures that have been brought in. She is slight but wields a kind of power from her hips when she walks, as if she has no intention of slowing down. I am ushered through to the kitchen, dripping, before she scoots upstairs to find me a pair of jogging bottoms to change into. While I wait, I look out at the garden. Newly washed leaves shine in the sun that has appeared with impeccable timing. It's early in the season and yet Clair's garden is full and growing; clouds of yellow brassica flowers, feathery nigella foliage, the nibbled white flowers of wild rocket. In the middle of it all, a square patch of broad beans, half in flower, half podded. On the left-hand wall, one apiece of a plum, pear and apple tree, the latter in bloom. Framed by the tall sash windows of the green kitchen, it is lovely.

We sit at a kitchen table that Clair tells me has held dozens of meals and bowls and dishes and top-ups before me and she presents a cake from a Tupperware, made to share with a garden volunteer who hadn't been able to make it that morning. 'So!' she says. 'What's the angle? What do you wanna know?' Clair was raised by an Italian family in New York's Greenwich Village – 'I always joke that I didn't see a tree till I was 17' – and her accent still clings to those etiolated vowels. A pragmatic kindness sits below the surface as she tells me about her life: two master's degrees from Oxford and LSE, a career as a primary and nursery school teacher, a deployment of her gardening knowledge – scant, self-taught – with the intention of bringing together local parents from different cultural backgrounds. Every few sentences she barks a kind of animated, yelping laugh, releasing a massive grin between her neatly bobbed curls.

Like other people I've met, Clair remains fascinated

by her grandparents, 'peasant farmers, based on my father's side, from Puglia – very rural'. She emulates their connection to the land, been chasing that culinary history for a lifetime. While her childhood was against grit and brick and concrete, pots of marigolds on fire escapes in a constant, tiny battle with the fire inspectors (a first rebellion, perhaps, against the authorities), her father introduced her to 'the sort of vegetables that are very trendy here now. I ate zucchini flowers when most people in this country weren't even born!' Clair can still recall spotting rapini or broccoli raab on a menu in an Italian restaurant seven years ago; when she asked where they sourced it they admitted they flew it over. 'For me, I think the gardening comes almost after the food,' she tells me.

It also came relatively late in life. Clair mentions her children in passing. She has two, and when they were small the garden was home to a sandpit and a climbing frame. As they grew up, she started to grow vegetables. Now, the garden offers food for herself and her partner, Ade, who seems to offer gentleness to Clair's vigour. I am let in on the minutiae of the water supply at RHS Hampton Court Palace Garden Festival and the perfect plot that Rainbow Grow never managed to land. She dramatically tells me about the plum glut that had to be harvested while she was travelling, and the futile endeavour of trying to grow chickpeas. Her horticultural approach,

Clair is insistent, is fundamentally in opposition to the British way of doing things. 'Maybe it's my Mediterranean soul coming out. You know, you grow this and you go out and you're cooking and you take a bit of this and there'll be a bit of that and you know. It's important to use the whole plant.' We have been talking for more than an hour and I have heard about many things, but I feel no closer to knowing much of who Clair really is.

In the garden, though, I can see her better. The puddles and the speeding cars that had rankled her so much on the walk to her house are only a few metres away, but feel distant. We walk around the beds and she speaks about what has worked, what hasn't, which trees were gifts and when she sowed what. The patter of identifying perennial and self-seeded annuals is affectionate and infectious – both of us sharing stories of what is growing in our spaces, what the names of things are. She is somehow softer; less businesslike, more excited. These are the beds where her food comes from.

I tell Clair the garden looks like a beautiful rebellion. At this time of year, food growers' plots are neat rows of potential, the beginning of the harvests to come. Much of Clair's has been left to go to seed. As I walk around I am torn between thinking about the bitterness that will arrive in flowering parsley and how much I want my garden to look the same.

Quite by accident, I've visited Clair at a time when my

feelings towards my own garden are shifting. It has been a refreshingly wet week, interspersed with bright, new sunshine. The garden is becoming an active, waking thing: clematis petals smother the ground, the poppies, calendula and cornflower seed I've scattered are turning into seedlings. Dormant perennials I'd forgotten I'd planted are rising from the ground with a vengeance. Swiftly, the beds are becoming more green than brown. Each morning I head out a little earlier, a little sleepier, to let myself be woken by the heady smell of rain-watered soil.

To see such growth is galvanising – and freeing. For the first time since we moved in, nearly a year before, I feel able to shake off some of the preciousness that has dominated my time in the garden. I've been scared of it, I think; worried I'd get something wrong, not make it look nice, watch things wither and die. But that striving, the determination to achieve something impossible and undefined, has stolen the gut feeling I rely upon when I garden. I'd wanted something designed and impressive, something cohesive and smart. In the process I've lost sense of my own gardening personality: the slightly slap-dash enthusiasm that has always powered me on, from planting those first doomed herbs in tomato cans on a balcony, seven years before.

In letting go, new desires appear: namely, to grow things I can eat. I want the garden to give me more than just distant beauty – I want to feast from it, to eat its produce, to take it into my body and taste it on my tongue. I plunge tomato plants against the back brick wall; I turn the right-hand bed into a herb plot. I put courgette seeds in pots, plants in the old tin bath on the patio. I push a handful of nasturtium seeds into the ground and make plans for wild rocket and kale and leafy things. Best of all, I don't think about whether they may fail, or be pecked to ribbons, or munched by slugs. I just want to invite in the mess. Increasingly, the garden that I have been tending to by myself had become a place of others' growing. Seeing Clair's has helped me to break the rules, to embrace rebellion, to listen better to what the garden can teach me rather than worry about how I can shape it.

Nothing comes out of the earth without having been placed there first. Matter rots down into nutrients, seeds are scattered, rain falls. The power in a handful of soil is hidden and dormant until the conditions – the pressures and alchemy – are right for growth and change to emerge. Anne's garden was inspired by a woman's life cycle, but it also reflected her own: the herbs grown through 40 years of learning, a life lived by the same power of determination that spurred her to travel as a teenager, to isolate herself on an island in her twenties, to build a 'powerhouse of women' on her land in the decades that followed. Clair had been a literal teacher, and the energy inside her that propelled her to be a community leader, to make space for marginalised people, was still potent. It's why she cared so much about the potholes outside her house.

But these women had used their power to go to the ground, instilling it into the earth so that it could nourish others by being passed on. We can grow things to feed people, we can grow to heal them, but Anne and Clair showed me that we grow to teach, too.

In the garden, a shift is happening. I am no longer the woman who spent a season staring at sullen, bare earth. The composting, the sowing, the planting, the plotting: the results are unfolding before me, and for all my picturing of what it might be, the reality is something I could never have imagined. Together, we have unfurled. The garden's unruly summer beginnings show me that beauty lies in the unexpected. The colours do not match, some plants are more boisterous than others, there isn't much here in the way of 'design'. Instead, I have grown a garden moti-vated by whim and curiosity, and now I am beginning to see what that looks like. Whim and curiosity are what have taken me across the country, to other women's homes, to their retreats and their favourite park benches. I'm beginning to see why, I'm beginning to see what I'm learning: that to be a woman is complex and powerful, something that is learned in the doing, and that our stories need to be louder.

8

BLACKFRIARS

PERHAPS IT WAS BECAUSE I started to garden on balconies, which are usually issued with typewritten leaseholder agreements about plant pot regulation, or started to garden at a time when it seemed like people my age should be doing other things with their youth, but gardening has always felt like a rebellion to me. In the beginning, I didn't know the full extent of how political gardening was: I hadn't considered the weight of privilege I carried in feeling comfortable to plant something in the earth, in the ease with which I could get soil under my fingernails. I hadn't acknowledged what permissions I'd taken for granted in being raised in the countryside, with a body that could run and play and dig there without incurring any questions. It took me decades to find the ground, and the release of gardening it, but when I started my identity didn't stop me, because in this country society has always accepted that white people garden.

The more I learned about gardening, the more I realised that this was a world in which I could move freely and that was a privilege afforded to only a few. I felt something of an outsider in the horticultural world; I snuck into the RHS Lindley Library without a membership card; I couldn't tell you the difference between Chatsworth, Gravetye and Sissinghurst because I'd never been to their famous gardens; Wisley left – and leaves – me cold, in all honesty. I hold no gardening qualifications, I don't understand Latin taxonomy, I endlessly second-guess my own technique. I still position myself outside the horticultural establishment, despite being given access to the Chelsea Flower Show and asked to write about it.

But, crucially, nobody interrogates my spending time in these gardens and institutions; nobody openly doubts my belonging to being outdoors with a trowel in hand. My body has never known what it is to feel unwanted, or out of place, on these grounds. My ancestors were the ones who claimed others' land for themselves, pillaged their plants and profited from them. Nobody made my forefathers farm or grow under enslavement or on stolen ground.

To have a garden is one privilege, to have the time, means and confidence to tend to it is another, and to ignore the political structures that have enabled both is a third. Our gardens can be havens, places to celebrate and grieve, to find comfort and relaxation, but to deny the privilege of this is to deny their power. To rid gardening of its rebellion and resistance is to dull its purpose. It's been years, and I still find the term 'gardening' an ineffective word for what I do outside – sinking my hands in the earth, watching root and stem curl from cracked seed casings, finding something new has bloomed in the early hours of a summer morning. It has so many twee associations, of neatness and nicety; a prissiness that feels deeply removed from the thudding compulsion that draws me outside, of the sex and death and life on show in every growing thing. When we garden, we change how a small part of the world works. The decisions we make – whether to spray our lawns, or not; whether to garden for the ecosystem, or not – have impact. Of course our gardens are political spaces.

I came to learn this mostly through the generosity of others who were less fortunate. As a child I roamed the village we lived in, with unthinking access to land and ease in being on it, but when I grew up and lived in cities I saw that land was something that was partitioned up and parcelled off. Public footpaths were less visible among the tarmac and the concrete; instead there were other boundaries, of postcodes and development plots, of private gardens. There were those that ran more deeply and more painfully, still. The ones I had to interrogate my own privilege to acknowledge: of class, race and belonging. I learned that parks had radical histories of political protest and petition, of a determination that all people should be able to enjoy green spaces, not merely those who can afford it. My favourite green spaces – the community gardens I volunteered in, the estates bursting with tree ferns and vines – had been saved by squatters and made beautiful for the greater good.

When I was starting to ask why women grew, a louder reckoning was happening: a new civil rights movement in the form of Black Lives Matter. BLM wasn't new – the movement had started seven years earlier after George Zimmerman was acquitted of murdering 17-year-old Trayvon Martin – but it had gained a new, global resonance after the killing of George Floyd in May 2020. By June, protests were taking place in defiance of a pandemic that was disproportionately affecting people of colour, with people taking to the streets to demonstrate for justice and equality, to demand to be seen. Institutions started to interrogate their own racial biases, to examine the extent to which they upheld white privilege. In the UK, the horticultural world largely panicked. The cathedrals of botany and plant science in this country are built on plunder, colonialism and enslavement: botanic gardens were established as a means of nurturing and developing plants stolen from colonised countries to grow industrially, often with the use of enslaved and exploited people, in other colonised countries. Many of the plants in our domestic gardens, even now, have whitewashed histories – of being 'discovered' by the coloniser, wrenched free of their indigenous name and use, and plunged into British soil as trophies. Members of the horticultural industry will still use, and own, the term 'plant hunter' quite merrily, seemingly unaware – or unbothered – that, traditionally, it meant to pillage.

In the same month that George Floyd was murdered, a study by Ordnance Survey found that one in five households in London had no access to a private or shared garden during the pandemic, with one in eight going without nationwide. A survey by Natural England established that Black people were four times more likely than white people to have no access to outdoor space from their homes. Parks and public spaces took on a meaning that felt both novel and centuries-old: that they were vital, a precious place to connect with the outside world.

Conversations about access, inequality and racism entered gardening's often sanitised mainstream in the midst of a harvest buoyed by a warm and sunny spring. Something sparked and ignited on social media, where people – often those of privilege, who had rarely had to think of their gardens as something that had also come from good fortune – refused to acknowledge that their gardens held the heft of politics. That among the bedding plants and the cut flowers, nestled among the vegetable beds and the rows of beanpoles, was an often-ugly history and legacy of who was able to grow, and how, and why.

I am still learning the politics of gardening, just as I am still learning about its plants and science and seasons. One of the reasons why many of us grow, I think, is because it is a constant education. Few years repeat one another precisely; there will be good seasons and bad, crops will fail and others will succeed. If our gardens grew the same way every year they would become invisible to us; it is these endless tiny changes that keep us captivated and curious. The more I learn about what it deeply means to garden, the more I appreciate its beauty: not just in the aesthetics of what appears from the ground, in the colours and shapes, but in the act itself. How brave and how brilliant to make a change in the world by making a choice in the ground around you. How potent to know that in the physical acts of gardening – of moving your body, smothering mulch, scattering seeds – you are contributing to something bigger than your back garden. How much more it all means with this knowledge than without.

But if you have fewer hurdles to clamber, it is easier to make change. I took on the garden – our garden, our first garden – with a heavy awareness of what it was to have it. It is small by national standards but large by those averages of the city. Beyond that, we are two middle-class white people in a Black neighbourhood that is being rapidly gentrified by people like us. This space was an extension of the privilege I had been granted throughout my life. I wanted to tread softly here; I wanted to do good by this plot, learn its history and its surrounds. I wanted to leave it better than I found it, not just for my own benefit – of having plants to admire and flowers to cut – but that beyond myself. I knew that there were women who were shifting the balance of power and privilege with the work they did on the earth. I wanted to meet them.

I'd met Carole when she turned up on my doorstep, shortly after we moved in. Through social media, she'd recognised the new neighbourhood we'd moved to – she grew up in the estate that looked over our garden. In a snatch of shade, we drank cold coconut water as she shared her memories of Brixton in the seventies and eighties and spoke about the documentary film she'd contributed to about the neighbouring Angell Town Estate. We kept in touch: Carole's the kind of person who makes things happen. She advises and creates, she works with arts initiatives and sets up charities to get young Black people involved in the outdoors. When I asked if I could interview her, she agreed. We meet on a brisk Saturday morning on her estate, a short stretch from Blackfriars Bridge.

When I first met Carole she'd arrived after a long wait at the barber's, where she'd had her fade seen to. Now, she's a hat tugged over it, a mask pulled up across the bridge of her nose. Her glasses dim and brighten in response to cloud cover. So much of her face is covered, and yet I'm in no doubt of her expressions – variously uplifted and opinionated, feverish and watchful. She carries a level of fizzing energy that feels like it should belong to someone taller. She knows how to keep bees and does so in a bright pink beekeeper's suit. Her stories meander and overlap, they swerve between history of what has been and intention of what is to come. More than growing, more than activism, I get the sense that Carole loves to walk, tracing the often overlooked and untold stories of the people who have shaped the city she was raised in as much as her own life here, a schoolgirl who fell in love with plants through roaming the rose garden in Kennington Park; a teenager who documented the uprisings of Brixton and the Domino & Social Club on Coldharbour Lane; a woman who has put her own stamp on these formerly anonymous, barren city plots.

We start on her estate. Carole rattles through the beds at the shared garden, according to who maintains them. 'So we've got Moroccan heritage, Jamaican, English – Northern, always telling me that, I'm like, "Yeah, I'm a soft Southerner, I get the message. But don't try me though." You have American-English and we have me, who's Ghanaian, Syrian, Jamaican. We have North African-British. We have Croatian. We have Spanish-English. We have Russian-Spanish. We have Portuguese. We have English. We have South African. We have Canadian. That's us. That's us in here.'

The garden is 10 years old, one of the many growing spaces in the surrounding square mile that Carole's set up over the past couple of decades. They are easily missed, tucked as they are between housing estates and high-rises. I've cycled through this snug triangle between the river and the railway countless times, but Carole could be showing me a corner of a wholly different city, one layered with music and protest, art and social housing, battles with building developers and the kids she's taught how to garden.

The Peabody garden is a smaller triangle within that triangle, an awkward bit of tarmac that Carole, as voluntary garden coordinator, has overseen since its creation. It's a role she speaks of with a mix of pride and begrudgement; since she moved to the estate a decade ago, Carole has helped to transform it, petitioning the council for new trees and planting, for play equipment and signage. She mutters about the drama on the neighbourhood WhatsApp group; I suspect she doesn't take any nonsense among the community gardeners: 'I say, "You just come, you do your thing, as long as you don't have a punch up. I don't want to hear about it, I could do without the aggro."'

There are benches and boxes of outdoor games; sunshine filters through the petals of the narcissi that line the paving. We are within hearing distance of one of London's main traffic arteries, helicopters chop overhead, but the green spaces here make all that feel distant. 'It is gorgeous,' Carole admits. 'Even in this weather, when it's howling down, just to be able to come out and have a potter and see what's in people's plots and go in the shed. It's incredible.' In the warmer months, Carole likes to eat her breakfast in the garden or the shed, waterproof thrown over her pyjamas, wellies on her feet. Her plot has a giant kiwi plant, with which she is extremely pleased. 'It's my favourite place to stand,' she says. 'I just come in here quietly, and nobody can see me.'

Carole says her love of gardening came from her mother, who grew up on farmland in Jamaica, raising cattle, chickens and goats. As a child in a houseplant-filled flat in Kennington, Carole would hear stories of her mother's life on the land, where she would go walking, swimming and fishing; the names of her goats and what happened to the dog that killed chickens. 'My love of the land is from my mum, and growing things,' she says. 'She told me the history of food.' During the summer holidays, Carole and her brother would be given the house keys and a packed lunch and let out to roam South London while her mum was at work. They'd pick blackberries in Lambeth, wander as far as Bermondsey. Carole's grand-father, too, was a grower who kept two allotments in Lewisham after moving to England from Jamaica. When Carole would take him to his hospital appointments in his final years, he'd make them stop by his old plot to complain about how standards had slipped in his absence.

It was only after he died that she found certificates he'd won for Allotment Holder of the Year, three years running. 'He just kept it quiet,' she says. 'It was incredible.'

We walk around the estate, Carole's stories interjected with asides at the housing association surveyors who apply glyphosate to the ivy, much to her umbrage, to the snazzy-looking basketball court where, just the previous year, Black teenagers had been issued with dispersal orders by the police simply for being outside their homes. It's a new thing, Carole says, this heightened police presence around children and young people spending time outside during the pandemic. 'They're rooted, like the London plane trees. They're like the furniture and the fittings, because that's where they live,' she says of the youth on her estate. 'Where else are they supposed to go? All the youth clubs are closed down anyway, even before Covid we lost three youth clubs in this area.' She has set up initiatives to help them feel more comfortable, more legitimate, in the outside space their city offers; having welcomed these young men as boys into her gardening club, she has earned a level of respect. They help her build planters and window boxes. Carole frequently challenges the police – she tells a brilliant story about shaming an undercover policeman for being given away by his 'appalling' outfit. 'When I watched *The Sweeney* when I was growing up, they had better clothes. Come on.'

She shows me the smart brick paving, the table tennis

courts, the outdoor gym, rattles off the funding applications made to achieve them. She clucks at the planning officers who work for the developers, calls them turncoats. In a flower bed sits a memorial plaque for the residents who have passed away. Above, a run of yoghurt pots on a ground-floor windowsill, filled with earth and seeds and lolly-stick labels. They belong to Peter, Carole explains, a retired teacher. She buzzes herself into his block and leaves packets of seeds on his doormat. Her love for this estate – which is handsome in its Georgian symmetry – is infectious. I tell Carole I can see why she likes it so much. 'Yeah, I love bopping about, chatting, putting up the posters,' she says. 'But I'm not doing it by myself. I'm old people now; I'm 54.' She says she 'uses sympathy' to get help with her projects.

The tour of Peabody is complete, and there's a lengthy, if invisible, itinerary ahead of me. Carole shows me Brookwood Estate, originally designed as part of a Fritz Haeg exhibition at the Tate Modern over the way 14 years ago, but looked after by Carole ever since. She's recently won funding to turn it into a forest garden to make it climate change resilient. The land in this postcode, she explains, is deeply contaminated – decades of factories, prisons and cholera have left their mark. Every garden Carole has made is on soil that has been brought in from beyond this postcode. We keep walking and stop to look at her favourite tree in the city. 'Look at the shape of this, look at the form of this tree,' she demands of me. 'And when it's got leaves on? Oh, it's to die for.' Carole tells me the story of a girl who loved the tree too, who came to her garden club at Brookwood. 'One day, I saw her behind the Old Vic, sitting on a wall with her mate with a beetroot in her hand,' she begins the story. The girl – about 15 then and, Carole says, 'a goth' – had been out for hours with a beetroot she'd harvested that morning, showing it off to her friends. 'That just made my day,' Carole says, looking up at me, smiling beneath her mask.

We're en route to the first community garden Carole came across – and the first she worked in. It's a sweet little space, clearly well-loved, with a pond and low brick walls. A wooden arbour sits over a bricked path. Anonymous, glossy modern flats rise up on three of four sides; the last one boasts a redbrick maisonette block where Carole's grandparents once lived. When she found this space 15 years ago it was home to the last pre-fab housing in central London; there were four of them, squatted in by the people who would go on to form the Bankside Open Spaces Trust. Here they created the Diversity Garden, with financial help from the Bengali Women's Group. 'It's why Mrs Begum, and some of those original founding members of Bankside Open Spaces Trust, still have raised beds here.' Carole had moved to the area from a women's hostel in Lewisham and was walking past when she got talking to one of the women working in the garden. 'They said, "Do

you fancy a raised bed?" and that was my intro to community gardening.' For a couple of years she worked on her
plot with her nephew. When a community garden job
came up to work for Bankside Open Spaces Trust, Carole
applied and was successful. 'And from then I was doing
community gardening,' she says.

Carole's always done some form of community work,
she explains. Before the Diversity Garden, it was Angell
Town, the infamous Brixton estate that was regenerated
in 2001 thanks to the relentless activism of Dora Boatemah.
'I just rocked up to her office, asked if I could photograph
her work,' she explains. For the next seven years, Carole
bore witness to the struggles and changes on the estate
over the road from where she grew up, the estate my
front-room window looks out on. Dora, she says, ossified
her desire to work in the community.

Ever since I first met Carole, I've expected to weather
some criticism of the rampant gentrification that has
transformed the neighbourhood that formed her. As a
white, middle-class woman who grew up elsewhere, I am
part of the problem; I am prepared to hear her thoughts
on this, even if I know I don't have any defence. I feel
deep guilt over what people like me are doing to areas
those who were here before had considered their home.

When gentrification does come up, though, Carole
blames Thatcher instead: her own mother bought their
council house. 'She was adamant she didn't want to own

property; now she's ended up owning two, you know, because what do you do? You have to be realist.' Her chat bends back round to the Diversity Garden, which we're still standing at the edge of. Its origins are embedded in her; she went to the funeral of the last pre-fab resident who lived here, visited the grave a few months ago on one of her sprawling walks. They wouldn't have been given this much space by the council now, Carole says; decades ago, Bankside wasn't the desirable area filled with bougie hotels and riverside flats it has become. 'Well done them,' she concludes firmly, 'well done to them to stick to those negotiations.'

I ask where Carole thinks it comes from, her relentless desire to fight for green space in a city so determined to endlessly build and stretch and sell. In response, she mentions her father, says that he was 'a proper activist' who lived with anti-apartheid activist Steve Biko. 'I don't have to look far to know where the kind of "let's get things done" comes from,' she says. But Carole's long made her work her own. Her views on social housing are complex. 'Why do you think living here makes you lesser than?' she posits. 'But also, equally, why do people have to just be giving you things? Why do you just expect, not just me and my time, but why do you think Peabody owes you anything? It's a very patriarchal view.' For all of the landscaping she's fought for, Carole admits that the estate can be challenging – there are people screaming in the early hours, there are certain buildings with residents that scare her: 'You just want to go and put your leaflet in and run back out very quickly.' Carole's work, making the play area nice, getting new trees planted, is part of a wider, very practical ambition: 'We all have to share this space. So how do we make it? How do we get along in this space? What bits and pieces can we do for ourselves without asking permission?'

She describes her determination. 'I just got this little fire that goes, "I'm not having this, how does this work? Nah, out of order, we've got to come and do something,"' she laughs. 'We must do better than this. We must never settle for second-best.' The whole point of that fire, Carole explains, is that it catches and spreads: she wants other people to give back to the community; she wants people to know they can rely on each other, to improve the space that they share. 'I have that spirit. But I will do this, and then I have to move on and leave people to do it. It's like, what are the young people who are working with me going to take away from this process? Because I know what I'm taking away. My intention is to step back, and back, and back.'

We end up where we started, in the community growing space at the edge of Carole's estate. There are new plans afoot to improve it: new benches, better planting, a gallery space near the fruit trees. 'This is what fuels me, really,' she says. 'Just to come in those gates and know that this was tarmac with dirt over there and a bare wall. And look at it, it's beautiful,' she tells me. I do look, and I can't disagree with her.

These are the kind of gardens I fell for years ago, the gardens that presented themselves as a language I wanted to learn. The ones pulled together with team effort from wooden pallets and milk bottles, the balconies crammed with pots, the proud window boxes that brighten up a block of flats like a slick of red lipstick. Still, now, they are my favourite. I remain poorly versed in the famous gardens – the ones nurtured by aristocrats, built with the spoils of empire to decorate country estates – but I can tell you about the tree-sized, oak-leaved pelargonium in a front garden around the corner; the lily-of-the-valley smuggled into the flower beds of my old estate by a guerilla gardening resident; the tiny community gardens of South London, squeezed into parking spaces. I see stories in these gardens, in the metal boxes filled with grasses and alliums; I see good intention and hope where there was once just pavement.

When we moved, I mapped my new neighbourhood by the front gardens on our street, marvelling at the container garden at the end of the road, where brassicas and vines and dahlias grew, and escapee oraches twisted under the front gate, still in bloom even in the depths of December. I found the community garden tucked to the side of Myatt's Fields. I spoke with neighbours who were planting up troughs with seedlings and optimism, and took a spare courgette plant from them with the promise that cuttings would be given in kind. I thought about ways of giving over the sun-baked plot at the front of the house to someone who would tend to it – my own way of sharing, of offering a space to someone who didn't have one – but got caught up in neighbourly legislation and concerns about the passing traffic. It all felt quite pedestrian and private: an entry-level beginning of gardening with consciousness. It didn't feel anywhere near radical enough.

•

It's a still, grey-skied day in North Norfolk and I'm looking out at neat, puffy lines stretching towards the horizon. This is a flower farm, modest by most standards at exactly an acre, but nevertheless part of a revolution. The flowers – at this time of year, tall drifts of the deep red *Sanguisorba*, aubergine-coloured scabious and starbursts of bright pink echinacea – are grown sustainably, with a careful eye on watering and no chemical input. According to a study carried out by Rebecca Swinn at Lancaster University, a bunch of these contain five per cent of the carbon emissions of a bunch grown industrially, or overseas, as the vast majority of the flowers bought in this country are. Flowers, and flower-growing, have an environmental impact that would horrify most who admire them: a bunch consisting of five imported Dutch roses, three Dutch lilies and three stems of gypsophila grown in Kenya (a major source of the cut flowers that end up in our homes) has a carbon footprint of 32kg – more than half that of an economy flight from London to Paris. It's also an industry that has long attracted a reputation of frippery – a nice thing for ladies of leisure to do. It's an assumption that straddles the cloudy gulf between fact and opinion; like any farming, flower farming is hard work. But the vast majority of small-scale female flower farmers registered with Flowers from the Farm, the main association of flower growers in the UK, own the land they grow on, which lessens the pressures to make a profit from their harvest.

This is Cel's farm. She's been growing here – on rented land – since 2013. Cel is short and smiley and robust. I spot her waving me down from the side of the A-road that buffers the land. She's in a workwear jacket, her tight curls cropped at her cheekbones. There's a sign for strawberries by her head, a hangover from when this plot used to grow organic fruit for the smarter hotels that line the nearby coastline, decades ago now. The farm is a neat square framed by trees and unfolds from Cel's workshed, an attractive structure painted dark blue with an elegantly rusting panel on the side that quotes Monet: 'I must have flowers, always and always.' There's a workroom for flower conditioning and seed sowing off a smaller, more open platform with a table and chairs, where we sit. The radio burbles in the background, until Cel switches it off and passes me a drink. Jewel-coloured bouquets wrapped in brown paper and bunches of lavender sit in buckets on the floor; euphorbia and lavender hang from the ceiling. The air is thick with that raw, good smell of pollen and cut stems.

Cel grew up in East London and you can still hear it, lightly, in her vowels. She was finishing school when work began on the M11 link road '10 minutes from where my mum and dad lived, and directly alongside the playground of my high school'. In the early nineties the site attracted road protestors. Cel's father is Guyanese, her mother from North London. Growing up, she says she wasn't exposed to nature much – 'we didn't go on walks in the woods because we didn't live near any, we had no car, there was no extra money to go and spend the day travelling some- where' – but meeting environmental protestors on her doorstep 'opened my eyes to a whole world that I'd not experienced before'. That world was one of food growing and local food systems. 'It was rare to find anything organic in a supermarket at the time,' she says. 'I was just experi- encing loads of different things.'

She was also 'pushed to go and study' by her parents, steered towards taking History of Art at university rather than heading off to art school as she wanted. 'There was always an expectation that I should go to university and stuff,' Cel explains. 'But I went there and it just felt like nothing was real. All of it just felt like a waste of time. So I dropped out of university, which was horrifying, and ran away to go and live on an organic farm in Kent.' Food growing was of particular interest to her because, she says quietly, 'There were times that we were hungry when I was younger. And I think one of the things I was thinking at the time was *why* should people be hungry? We've got all this land in this country. Nobody should be hungry. Nobody,' she repeats, 'should be hungry.'

The process of 'actually taking the seed and putting it in the soil, nurturing the plant, and then being able to nourish yourself from it? That whole process, I just found fascinating.' Cel went on to study garden design and organic horticulture, visited gardens for the first time and volunteered on a stand at the Chelsea Flower Show. 'It just opened my mind up to a completely different world.' She also met her husband on the organic horticulture course. 'We're still here years later – he always says I made the wrong choice to go to college that year!' She laughs, fondly.

Their two children came along, and with them camping holidays and walks in the woods, a garden design company and a landscaping one. The couple realised they didn't want to raise their children in London and moved to North Norfolk, where they'd always enjoyed holidays. It was meant, Cel says, to be a simplifying of their lives, but for years her husband was still commuting down to

London, while Cel took what work she could in a local garden centre. When she took the lease on the farm, it was in disrepair. Eighteen months of gruelling graft followed: days spent clearing, nourishing and growing in the overgrown beds, nights spent stacking shelves in Tesco. At the weekends, Cel and her husband would take what flowers she had grown to farmers' markets on either side of London. She must have been permanently exhausted, I say. 'Yeah, that did take a lot of work,' she says, simply.

Cel shows me around the farm. From the workshop, the field is beautiful: soft rows of colour, a haze of insects above them. Up close it's more of an engine room; there are walkways and stakes, plots separated by season and need. There had always been a plan with the farm: to sell wholesale – to local florists, mostly – within five years. Beyond that, to be able to own land to sell on. There's an obvious practicality to this: Cel has to make rent to her landowners in order to keep those years of hard work viable. But what she's part of, what she is striving for, goes far beyond this acre. 'The global floriculture industry is horrifically detrimental, not just to our natural environment, but to people,' Cel calmly spells out. 'Not just the treatment of people, but their exposure to chemicals. Then you think about the wider implications about this global trade. That is the carbon footprint, which is unbelievable, which has those longer-term impacts on our future.' She pauses. 'There has to be a sustainable option. We are always going to buy flowers. If you want a sustainable option, you need to be looking at those small local growers to provide seasonal, sustainably grown flowers that are not covered in chemicals and are catering for your local bio-diversity as well. We want to be doing that for the future. But we can't do that if we don't run profitable businesses. Because the truth of our world is that we live in a capi-talist society, and I have to own the land to grow on to have that sort of business, which means I need to make money. If your business isn't financially sustainable, it won't be here for the long term; if we're not here for the long term, who's providing the sustainable option?'

I ask Cel what it is that drives her beyond this deter-mination. This is not an easy way of making a living: what stops her from giving up? 'I just want to grow plants,' she says. 'I really identify as a grower. I'm not a florist. I don't do floristry, I don't care about making arrangements. I want to grow plants. I like when the crop is ready. I just like to go through it. Cut a bunch in tens, they go out the door, I get paid. But then I can just carry on growing.' She stops for a moment, and her round cheeks rise: 'Finding newts in the compost and having to rehome them. Cutting the lavender. I can just take a minute, whilst I'm cutting and just breathe, feel the sun on my face. Can't do that if I'm sitting on a checkout, or if I was in an office. I want to be out doing this. And I need to hold on to that thought about sitting in the sun with the lavender in the middle of January, when the wind is howling, and the rain is like ice on my face. And I'm hauling barrows of compost onto the soil to get things mulched. I'd rather be outside on a freezing cold winter's day than stuck in an office.'

•

I had been trying to put these women who grew to make the world better in a box, one vaguely marked 'political' or 'socially conscious'; I had been wrangling with my own myopia and propulsion in growing. In the midst of climate catastrophe and civil rights movements, when gardening lets its privilege show so easily, whatever I did in my garden felt naive, an indulgence. I wanted, I suppose, to hear from those who gardened because they saw it as a way of challenging larger societal structures. I believed that was the impetus: that going to ground enabled grass-roots change. What I found instead was a compulsion. One to do better, to make better, to put back in the earth matter that humans have eroded away, to create safe spaces to incite change in a world that incites so much rage and violence. But that compulsion also came from the need to be outside and engage with the earth. To throw ourselves down and among the soil that we may have spent so long trampling on, ignorantly. When we stop and notice, we realise what we have – only then does it become far harder not to fight for its preservation. It always goes back to the earth.

I go to Kent to visit Sui, a gardener, writer and artist. In recent years, Sui's embarked on a new pursuit of trying to make the outside world she loves more accessible to marginalised people. She founded an Instagram account, Decolonise the Garden, to try and expose the legacy of colonialism in British gardening and the horticultural industry. She speaks on panels and events, upholds marginalised voices and interrogates the apathy of many major gardening and heritage institutions to be more inclusive. I've learned a lot through Sui's work, and I want to talk to her about it away from a phone screen or an online webinar. We sit on her patio, and as she speaks I savour the crispness of the magnolia leaves she's recently pickled, feel the gentle depth of nettle tea in my gullet.

'To me, the enjoyment of plants and growing and gardening is like food, or music,' she says. 'It's just so innate in all of us as humans, and has been since the dawn of human time. The idea that only some people would be interested in it based on a fake construct of race, it completely blows my mind. It floors me every time.'

We talk about race, about racism, about the structures we live in, about how those things interact with the outdoor world differently for different people. 'I cannot imagine what it feels like to go out there and to walk and to not have any inkling of what it would feel like for someone to see the colour of your skin,' she says, evenly. 'Or to question why you might be there, or to think that English isn't your first language . . . all those things that come from assumptions being made on the way you look. I have no understanding of what it would feel like to never have experienced that or to not know what that feels like.' Sui isn't telling me this for pity, but for explanation: as a woman of colour, she has never known what it is to slide through the world, especially that of rural England, without being seen as different – something I have been fortunate enough to never have had to worry about.

Sui decided to tread lightly in her garden almost imme-diately after moving here seven years ago. It's a gorgeous plot, the kind that invites you to wander. Crimson camellia flowers fade to pink, then mustard, on the lawn; a handful of narcissus are in bud beneath a swooning mature magnolia tree, blush-coloured flowers recently touched by frost. When Sui and her husband took it on, the garden had been dutifully maintained but hardly loved by its two previous custodians. She found ground elder and bracken, the bold spires of yellow archangel – tenacious plants often seen as weeds. 'I thought, I can either drive myself mad trying to control this stuff, and treat it like I had to anni-hilate it, or I could just garden around it and live with it,' Sui tells me, as she walks me through the trees, around the laurels and past the run of compost heaps behind them. The air is thick with birdsong. 'I knew I was never going to win that battle, so I just thought I'd have to get on with it.'

That pragmatism led to a new understanding of beauty. A few years ago, she decided to stop mowing the lawn – something that would take the best part of a day – and leave just paths in meadow, instead. It was an act of resistance against what Sui calls 'that overwhelming need or urge to kind of *do* something'. Instead of constantly cutting the grass, she decided to just do it once a year; to 'leave it and enjoy it'. Soon, marsh orchids started to grow alongside the grass, glow worms drifted against the summer dusk, slow worms turned up. This is not a garden, Sui tells me, that impresses upon first sight but one that rewards bearing witness daily. 'What it is about having access to a space like this is seeing it all of the year, and seeing it in all of the light, and the grasses are lit up,' she says. 'I don't think you can always get that connection through a brief moment.'

Sui, who has an open face and a neat, dark bob, speaks with quiet and considered conviction, when not gently chastising her dogs, who are excited by my arrival. She sees her role here not as owner or occupier but custodian – the garden surrounds her home, the oldest part of which dates from the 1750s. 'People have been working the land here for centuries. I always had that sense that I was just passing through here. I'm just looking after this space, and I'm going to enjoy it and hopefully do right by it while I'm here.'

Stepping into this legacy – of land caretaking, of gardening – is something she has discovered for herself, having made a career change and trained as a gardener after working in an office job in London. The training led to gigs in grand estates and manicured second homes; for a while, she worked at Windsor Great Park and lived in a worker's cottage on the site. 'It made me realise that you don't actually need much,' she says. 'I was in this tiny cottage, and it was the most beautiful place I'd ever lived. I had access to incredible beauty.'

As a child, Sui and her brother explored the Lincolnshire village they grew up in. 'We'd put our bumbags on and take a notepad and a pencil with us, go and chase butterflies,' she tells me, which instilled in her a strong connection with wildlife, if not an academic understanding. Her mother's family farmed in Hong Kong, but when her parents moved to England they took jobs in Chinese restaurants and takeaways; their shifts meant there wasn't much time to look to the small front garden the family had. 'I can't say I grew up in a family where they were all gardening, because I didn't,' she says, simply. 'It annoys me so much when people say "People of colour don't like gardening" because so many of those people don't have the luxury to garden, they may not even have a garden.' What pushed her towards retraining, she tells me, wasn't a desire to grow so much as a longing for meaning. The tipping point came when she saw a goldfinch on the terrace of her flat in London, and wasn't able to name it – nor the weed it had landed on. 'I was like, "How do I not know what that is?" I just thought I needed more of it in my life.'

Sui's social justice work didn't precede her gardening, and her attitude to her garden has changed over time, too. One informed the other. She's always been opinionated, she tells me, and had a clear and strong sense of her beliefs, but what has changed is that she now feels able to vocalise it. 'You can believe something in your gut, and then someone comes back at you and you don't know what to say to it,' she says. 'And I think I know the difference now. I know the words, I have the words, I understand what's happening and I'm not gonna let you push me over. And that feels so empowering.'

Sui and I talk for several hours, long into the afternoon against the squeak of calling great tits. Before I leave her, to head back onto the M25 into Friday evening traffic, she draws an invisible line of similarity between us. We both grew up in the countryside, neither of us gardened much as children, and then we both turned to the ground in our twenties, she points out. 'There's something in us that wants that connection,' Sui says. 'It's just inside you. And I think that's true for most people, it's just that you don't always get to peel back the layers to find it.' There was a lot of stress in her childhood, she says. 'I've always felt like I need peace. For me, gardening and being in the natural world gives me that like nothing else: one of the pleasures is that there are no people, there's no judgement.' How closely peace and politics can sit together, how much people want to divide them. Sui's garden galvanises her, but that doesn't stop it from being her retreat.

The stuff of what makes us – where we play as children, who tells us stories, the journeys our families have travelled to rear us – comes out in our garden. Carole gardens for her community, but that seed was planted with the stories her mother told her and the activism of her father. Sui's parents garden now that they're retired, but the fact they couldn't during her childhood meant she found her own way to the earth. Even when those who raise us can't show us how to garden, we sometimes still encounter them when we discover the ground ourselves.

I was beginning to understand the privilege I carried when I gardened, when I walked around the park, when I craved broader horizons, but I still didn't know how my heritage had affected my hunger for the ground. Perhaps, if I tried to trace it, I might find out.

9

BROCKLEY

A GARDENER WHO GROWS OF their own volition, without the insight of a relative or friend, is a rare find. Most of us who grow attach our interest to someone else, often someone older, who shows us what to do. Of the women I surveyed, a huge majority named their mother or a grandparent in response to the question 'What drew you to gardening?' In the conversations I was having, space was often made for ghosts – for grandparents and parents, for beloved aunts and sometimes sisters: older people who were passing on the skills they had learned, and with them, a passion for growing. The lessons were often informal – a hand-held tour around a grandparent's garden, an early bite of freshly harvested produce, the smell of cut flowers on a kitchen table – but enough to make a memory we chased in adulthood. 'A gardener's grandmother will have grown such and such a rose, and the smell of that rose at dusk (for flowers always seem to be most fragrant at the end of the day, as if that, smelling, was the last thing to do before going to sleep), when the gardener was a child and walking in the grandmother's footsteps as she went about her business in the garden – the memory of that smell of the rose combined with the memory of the smell of the grandmother's skirt will forever inform and influence the life of a gardener, inside or outside the garden itself,' Jamaica Kincaid writes, in an essay called 'The Garden I Have in Mind'. 'Memory is a gardener's real palette; memory as it summons up the past, memory as it shapes the present, memory as it dictates the future.'

Our memories do inform our gardens; we grow plants to remember people by, to mark particular moments in our life. In North London, Rhiannon and I spoke about her garden as a place of recovery, but I'd originally rekindled contact with her after I'd read a piece she'd written about finding traces of family members she never knew in her garden. Uncanny coincidences, like her husband picking her up a cheap rose from the Caledonian Road and it being the same obscure 1960s variety that her grandfather had loved, which became tethers to a horticultural heritage she only discovered after finding gardening for herself.

But Rhiannon also remembered her grandmother in her own garden, a woman who would take her by the hand when she was small and show her what was growing, treat her to purees made from the apples in it. 'She had a very difficult upbringing; she was born illegitimate, as they called it, in a mother-and-baby home and then raised by a horrible family who treated her very badly, and so she left home at 14,' Rhiannon said. 'I think for her gardening was probably a way of dealing with some quite difficult emotions that she had to live with all her life. Pretty much the only thing she knew about where she'd been born was that she was from Llandaff, and so she grew dahlia "Bishop of Llandaff" as a link to some roots, I suppose.'

We uncover memories in gardens, but they offer us a place to make sense of ourselves, too. I imagine these connections to be slippery and sinuous, twisting around one another. Like bindweed, they are difficult to root out from the gardens we make even many years later. In South London, the bare patch of ground Louise wanted to transform was shaped by the roses in her grandmother's garden that had captivated her as a child. For Louise, it went beyond mere memory: she and her grandmother were very close, and Louise felt that the older woman admired something in her granddaughter, who relished the freedoms that her generation were never allowed to have.

Marchelle, meanwhile, would visit the botanic gardens in Cambridge to feel tethered when she was homesick, having left Trinidad in her teens to study there. In the glasshouses the plants were familiar; she liked to breathe them in. 'It's something about the humidity and the smell that's so much like home,' she told me. 'It was just something I did instinctively.' Years later, in a garden in Somerset, she would trace her Anglophile grandmother in the roses that grow so much more easily here than in Trinidad, and connect to her paternal Chinese grandmother who kept chickens, by more deeply appreciating the offerings of the land. This garden, so far from the country where she was raised, still conjured connections to the women who raised her.

We don't have to be related to those we think of in our gardens. In North Norfolk, I sat in the back garden of Cosey Fanni Tutti, the artist, writer and musician known for her involvement in provocative seventies punk band Throbbing Gristle. She has always gardened, even while living in a countercultural commune in Hackney, and talked about the plants in her garden – which she transformed from a school playground 30 years ago – through the people who had given them to her: a pink rose named after her late neighbour Barbara, others from a friend named Les, a clematis from her sister and a cranesbill geranium from a woman she met while she was in hospital.

When I started to garden in my mid-twenties, I thought about my grandfathers. Theirs were the gardens I remembered most strongly from my childhood, even though I spent far more time in the one my parents grew. I was young when my grandmothers died, and have spent the years since building an understanding of the women they were through memory and left-behinds, dresses and postcards and photographs. My grandfathers loomed larger – the colours of their jumpers, the meals they cooked, the way they opened their front doors. And so, because they were gardeners, I believed my compulsion to garden originated with them. I traced my family history, found botanical artists and farmers, and tried to place myself among these strangers in grainy photographs.

But I was now beginning to question this narrative. My grandfathers' enthusiasm for gardening never overlapped with mine; they died before I started growing. I was trying to base an answer on childhood memory and hunch, on family story and snapshots. I'd last been in these gardens 5, 10, 15 years ago. If I wanted to know how they had shaped me, I'd have to go back.

In the last few years of his life, my grandfather went on a genealogical quest from what used to be the playroom. Against orange wallpaper, around the piano, papers piled up. My dad was taken along on this search through census records and birth certificates; strangers were contacted with word of a shared relative. The outcome was two lengthy family histories: one of my grandmother's side, the other of my grandfather's, both printed and distributed among the family. I took both from the shelf, traced the names of 19th-century strangers I was related to, looked at old photographs. In the final pages my grandfather had written about my grandmother, explaining that, due to family finances, she was 'unable to follow her first love, botany, in further education'. My grandmother had died when I was seven; I have memories of the richly coloured dresses she wore and the clipped tones she spoke with, the steaming bowls of buttery pasta she'd serve her grandchildren at family lunches and, in crystalline focus, the delight she took in my Polly Pocket collection. This, this throwaway half-sentence, felt like a hidden treasure: all this time, I'd placed an orphaned affinity for gardening in the palms of my grandfather, not knowing that there was a woman in my lineage who had been fascinated by botany since girlhood. I emailed Dad, clamouring for more. 'Gran loved botany/flowers etc.,' he replied, adding she'd learned it at school and clung to what she had, knowing which plants were from what botanical families. Weeks later, I visited a public garden with my parents and he mentioned it again. 'Gran would have loved to have known you were so into plants,' he said, as we walked through grasses that swayed above our heads. I said I'd not known she loved it, that I'd always attributed the gardening to Grandpa. 'Oh, no, he looked after it later, but she was the real enthusiast,' Dad replied. It was Gran, he said, who designed the garden, who carved out its image. One tiny revelation that felt like it changed everything. How swift I was, how unchallenged I was, in believing the men in my life were responsible for these things. How easily a woman's work – and with it, her wisdom, her fascination, her studies – is diminished.

For months, I'd been unable to let go of a desire to

retrace my childhood footsteps in the gardens I'd grown up in. Plants and the cycles they grew by had interrupted my early adulthood, seemingly from nowhere. I still didn't really know why I had started to garden, and the easiest – laziest, maybe – answer had been to say it connected me to my grandparents. Increasingly, I knew that wasn't the case; the impetus was fiercer and more complex than that. There was something deeply buried that I wanted to uncover. So many women I was reading about or speaking to were starting their garden stories with the ones they had encountered as children; it seemed that might be a place for me to start looking.

I no longer knew the people who lived in the house I grew up in, in the houses where my grandparents had lived. There were cold, dark weeks where I put off writing letters to the occupants. It seemed too simple, almost too whimsical a concept: that a stranger would pick up a handwritten envelope and find a plea inside. 'I grew up in this house, could I visit the garden?' When I did sit down to write them, I thought about the letterboxes they would go through. I wondered if the little black post box my father mounted on the wall would hold the envelope. I thought about the heavy velvet curtain that kept the draught out from the door. And then I posted the letters with stamps too expensive for the job and forgot about it.

A few nights later, after I had fallen asleep, Graham texted me. 'I received your card,' he wrote, 'and I moved to Ivy Farm at the end of last October.' It was a kind and formal reply; separated into neat paragraphs. It would be lovely for me to visit and hear about what had happened at the house I had grown up in before his time, he wrote. But he was concerned – the 'gardens appear not to have had much love for a while'; he was worried I would make the trip from London only to have my memories ruined.

I persisted. 'The bottom half of the garden was left semi-wild for most of my teens,' I told him, truthfully. 'Please don't worry about the state of the place.'

After sending it, I imagined all levels of disarray. Waist-high weeds and foxholes; the apple trees that occasionally bore fruit left to canker. But it never was an immaculate garden: loved, slightly feared, I think. It sprawled and held histories, a sweep of land that had huge promise for those who were able to get a grasp on the space. My parents made good of the things that had been left behind and ignored – the always-leaking pond that released putrid smells; the mounds of rubbish that elicited chicken wire and Victorian medicine bottles – and built beauty into its lawns. Still, though, it was a place that was maintained with a ride-on mower and bonfires. My dad would eagerly harvest the fruit from the apple, pear and plum trees – the latter in such quantities that I was put off eating plums for a decade – my mum grew herbs, sweet peas and red hot pokers within grasp of the back door. My best friend called it Eden. For me it was home, a place I took my stretching body out to in drifting late spring evenings and long summer holidays, restless and impatient for life to happen to me.

Graham replied swiftly. Within the hour, we were speaking through FaceTime; he was keen to walk me around the garden and the downstairs of the house. Really, I wanted to return there in person first, rather than through these well-meaning pixels in my palm, as I sat in bed on a Saturday morning. On my screen, the garden looked bigger than I remembered but otherwise unchanged, and Graham exchanged his plans for my memories as he paced the length of it, and then back again. Where he wanted to build a teepee lay the footprint of the marquee for my sister's wedding; where he planned an orchard I conjured the lazy summer wasps, drunk on perry. His was a generous act, and I was grateful for it. But the excitement was mutual: he hadn't been in the house but five months and I sensed Graham appreciated a captive audience interested in his new home.

We made plans for a visit in a few weeks' time, and I texted my family about the whole peculiar thing. Nostalgia is a powerful drug, and for a short while we wallowed in it. I spoke to my parents, watched their faces light up with reminiscent understanding as I retold Graham's story of walking into the house for the first time and feeling at home. Later, my dad sent over photos of the garden as they had kept it, and I saw a space I knew innately with a whole new understanding – of herbaceous borders and planting schemes I'd not only forgotten, but felt I'd never before seen. The vista was familiar, I could have sketched it from memory, but here were plants I recognised from my own, adult gardening experience rather than childhood memory. *Lamium*, *Digitalis*, *Scabiosa*: a language I'd taught myself with a grown-up tongue. These were the plants I'd bedded into containers on my tiny woodland balcony in South London, 15 years later; plants I thought I'd discovered for myself. Perhaps I had been drawn to them by some deeper pull, found them beautiful because they had bordered the paths my growing feet had run down.

My parents had left Ivy Farm eight years earlier; Graham was the second subsequent custodian of the place. A sprawling, hodgepodge house built across centuries, it was the kind of home people just looked after for a bit, rather than owned. My sister and I baked our adolescences into the inglenook around the Aga, the rusting colour of the lounge walls, the particular creaks and groans old timber makes. It was a building that trapped voices if you tried to call for someone, but I could always tell who was on the stairs, or crossing a ceiling, just by the sound of it.

Graham invited me to see the house but I didn't want to. My fear that seeing another person's furniture, another person's paint colour on the walls, would temper my memories outweighed any curiosity. Instead, he suggested I let myself in by the back gate when it suited me, one

Saturday morning late in April, warm enough to leave the jacket I'd brought tucked under my arm.

With that lift of the sneck, the push of the gate – stiffer, now – I inhabit two places rather than one: that of the now, as a woman in my early thirties, and that of the then, as a teenager returning from school. This was what it was to walk down the garden: to remember things I thought I had forgotten, to encounter a fierce wall of memory. Where I'd lain reading during the summer holidays, where we'd had family lunches and dinners at the picnic table, where I'd been sent out to cut herbs, where I'd stand and shout, and shout, and shout at my dad, temporarily deafened by the mower. The views – unspectacular, of open fields and the backs of houses – are things I've not thought about but are nevertheless the same. I can taste the listlessness and longing of my adolescence as clearly as I can picture my first mobile phone or smell the sweet hit of my first perfume. This garden was the space that held my loneliness in the wake of my siblings moving out. Time then was long, something to be waited out. That residue lingers with having come back; I question the purpose of my journey, of travelling here and seeing what might happen. I wonder what I should do with it, now I've walked up and down a bit, chasing memories.

I am seeing the garden as an adult, and an adult who gardens, against this reminder of my adolescent self. I mentally partition this long stretch of land into dozens of

London gardens; I marvel at the sheer expanse of it – much of it newly cleared, ground I've never walked on before. I spot yellow archangel and green alkanet; I catch the faded flowers of the cow parsley; I clock the wild primrose and fat bumblebees, hear the cry of a pheasant in the distance. Birdsong muffles the occasional thrum of a vintage plane taking advantage of a clear day, marking the soundscape of my childhood still intact. I enviously eye up bundles of pea sticks and wonder if I could take them home on the train. I whip up imaginary planting plans and fantasy designs on the space. I see more potential than I do loss.

Making my way back up the garden, I sit under the only apple tree still in blossom – white sprays of it, as much on the lawn as the branches. A siren goes by, and I think of my mother in the kitchen, her remarking as she hears it. They are uncommon in this part of the world. I had always thought of the garden as my father's space. He was the one who spent weekend afternoons out here, who built bonfires in blue overalls, who slowly chipped away at the piles of rubbish that had built up over the years. Decades on, and his work still stands: the beds are full of dark purple hellebores, one of his favourites; the mower shed he installed stands well, the pond he filled in remains grassed over.

And yet, sitting here, it is my mother I feel most strongly – or rather, her absence. I miss the schmozzle of pots outside the back door, the cluster of hostas around the sheds. The stones she picked up from our days out no longer decorate the patio; the bench she chose in honour of her parents has moved, the corner it stood in emptied. All of the things I loved about this garden – the towers of sweet peas, the Welsh poppies, the coldframe tucked behind the garage door – have gone, and they were her doing. I find it difficult to look at the beds beyond the back door – a scrubby mess, now – knowing how beautifully she kept them. I see her artistry, her cleverness, in the hard landscaping that still stands.

This garden didn't make me a gardener, but it urged me to find my own path. Maybe it was this green space that made me seek out others once I'd left for the cities I so craved, that instilled in me a permanent search for birdsong.

Graham catches me as I walk across the gravelled path to the gate; I thought I'd wander around the village. Instead, we meet properly and he lays out his plans for the garden – ambitious things, with good times in mind. 'I found something that I think was your mum's,' he says, minutes before my cab turns up, describing a hand-painted house number sign. I dismiss it, say she never created such a thing, as he emerges from the back door with a ceramic plate, handwriting undeniably the neat italic script of a former schoolteacher. He turns it over, reveals 'S. Vincent' carved into the back. Sheepishly, I claim it as hers, and cram it into my backpack. Heading back to London – a laughably easy journey in adulthood, rather than the grand adventure of my school days – I hold this revelation heavily, sad and ashamed that I never gave my mother credit for the beauty she surrounded me with.

•

'She had moths, and spun her own silk. The temperature in Cyprus: it was perfect. They had date trees in the garden, and so she would feed the moths on date leaves and then get their silk and use it to do her embroidery and knitting. It's very common in Cypriot households to have this one spare room that's the hottest room in the house, and there you'd keep moths, dry mint, dry out *mulukhiyah* – use it in some way to store or dry out food.' Şifa is telling me about her maternal grandmother, a story passed down by her own mother. In the weak sunshine of a London June she remembers summers in Cyprus. 'You know, you would just go into the garden and pluck figs from the trees, and grapes. It was very romantic,' she says, with a knowing smile.

Şifa's about my age, and has lived in a ground-floor flat in South London with her partner for the past five years. Her eyes are as dark and shiny as beetles' backs, and crease when she smiles; she's a halo of dark curls. We sit opposite one another in low-slung garden chairs, sectioned off from her neighbours, with whom she shares the long, skinny garden, by a wall of bamboo. Through large doors, I can see a living room filled with her brightly coloured art.

Unlike a lot of the people I've spoken to, Şifa has always gardened. She gardened in childhood, she gardened as a teenager and now, in her thirties, she gardens as an adult. She nurtures dreams of escaping to the countryside to keep chickens, growing what she eats and pickling and storing what she doesn't. Şifa hungers for the self-sufficiency her grandmother lived on, but she's also in the midst of a career change and a practising artist. Behind her lies a vegetable patch, where, after the trying, slug-heavy spring we've had, brassicas languish. 'Most of our kohlrabi has been eaten up,' she explains, a little resigned. Gardening during her childhood, which took place three miles over the rumpling hills of southeast London, was a practical thing, helping out her father while her mother 'pottered around' inside. 'We would be clearing and tidying, not necessarily knowing what each plant was, but we knew how to keep it alive,' she says. It was while she was living at home and studying at university that Şifa 'started to really actually love gardening'. The garden offered something 'therapeutic' when she took breaks while working. In her second year, she claimed a bit of her parents' garden – a space, she tells me, that is 'full of trees, and just beautiful' – to start a veg patch. 'There was this pathway at the side of the house that didn't do much but it had the best sun,' Şifa says. 'So I started growing like, runner beans and tomatoes and all sorts. Because the one thing my parents never really did, considering that their parents grew a lot of food, was they never really grew much veg. I started doing that and both my sister and I got really interested in it.'

I ask what her parents thought of this, the new plot Şifa tended to. 'They teased me about the fact that it wasn't enough, they were always like: "Oh, well, that's probably a dinner's worth?"' she says, laughing at the memory. 'But they really liked it. And it would make them share stories of their childhood and eating from the garden.' Growing to eat seemed to have skipped a generation, but when Şifa and her sister picked it up, it rekindled family tradition: she tells me her sister has taught her children to grow – 'they argue about who wants to be Monty Don!' – and Şifa is determined to teach any children she might bear to garden, too. 'I feel it's this skill that will get passed on and on and on,' she tells me.

Şifa never really had a relationship with her grandparents, she says. Cypriot on both sides, her maternal grandparents died before she was born, and she only met her paternal grandparents a handful of times. 'I think I just love this idea of reconnecting with my grandparents in a way,' she says. 'I would love to be like my grandma.'

More than a decade has passed since that vegetable patch Şifa grew during university, but she and her sister still tend it; her father is not as well as he once was. 'My sister and I now go in and do that garden for them when we can,' she says. Last year, he finally started his own vegetable patch in the garden, having cleared a spot previously occupied by a broken-down shed. Şifa took him seedlings and cuttings. Over the hills, a twin of her own vegetable patch appeared in her parents' garden: 'He basically grew everything I was growing,' she says. 'He sits down there in the full sun for the evening and reads, it's his favourite spot in the garden.'

Throughout our conversation Şifa mentions cuttings and plants: there's the *yeni dünya* trees, which grow in her parents' garden – 'you can tell a Cypriot house because they're definitely going to have a lot of y*eni dünya*s' – and the broccoli that grew so rampantly for her last summer; there's a jasmine that has been passed through her family, and the 'Mustafa Mint' grows tall and upright and fragrant around her garden. 'Mint is something very important to me,' she explains. Şifa cooks with mint and dries it; the variety in her garden has been passed through her family for generations, given to her father by a relative as a cutting. 'I've put it everywhere I've lived,' she says. When I leave, she hands me a pot in a plastic bag that we nestle gently into my backpack. I smell mint all the way home, cutting through the thick air of the dual carriageway, feeling the weight of a legacy between my shoulders.

Şifa speaks about the role of the front garden in Cypriot culture – 'it's where everyone sits, where the tables and chairs are, it's where you would sit for evening dinner and you'd talk to the people who walk by' – and it feels like her garden holds that same communality. Things to remember people by, things to eat, things to admire. Şifa is one of the few women I speak to who does not apologise for her garden, nor explain that, had I come earlier or later, I would have seen it look better. Instead, she seems to know it – its beauties and faultlines – deeply, for it is what she has made. 'There are a lot of things in life where you might feel a bit intimidated,' she explains. 'You'll see what someone else has done, or made, and covet it. But I never get that with gardening. Never, ever get that with gardening. I don't care if my garden isn't someone's aesthetic, it makes me happy and I get what I need from it. I don't feel that it needs to be perfect. I don't feel it has to be necessarily full of flowers. People seem to not like green so much but I love green. They don't think that's exciting enough. But to me it is.'

As Şifa spoke about her love of green, I started seeing her garden in a new depth. The different markings of the bamboo leaves, which stitched a curtain around her decking. The triumphant spires of 'Mustafa Mint'. The creeping, self-sown scabious and chives; the expectant foxgloves on the cusp of blooming. So often green is a background colour in a garden, relegated to lawn and anonymous shrubs. It can be easy to overlook, this building block of the things we grow, the stuff of cells and chemistry. When I got home, I saw my very green garden, which others had questioned, with a new appreciation. Still, months later, I think of Şifa when I think of how overwhelmingly *green* the garden looks. With one passing comment she has taught me a new understanding of beauty.

There's something in the building blocks, in the stuff that makes us and in the histories we bury and uncover. As much as Şifa's garden was made of greenery, as much as she had grown it in her own understanding of beauty, it was also the product of her lineage: of her father and his lessons, of the legacy left by the grandmother she never knew. I see invisible sinews of connection between myself and Şifa, between our gardens and how they carry a quiet longing for grandparents and people we have not known, or met, while holding a creativity that is all our own.

•

When we lose someone, we turn to the earth. We bury our dead, we scatter their ashes in their favourite places. We plant trees to welcome babies and we plant trees to mark our grief, too. But we also go to ground to connect with pasts we can't unravel, to find parts of our heritage that don't exist on paper or in others' bodies, to make sense of who we are. Sometimes the desire to garden can feel so out of place that the only way to understand it is to look back to the nearest green fingers we've known.

In Regent's Park, I meet Aimee, a horticultural apprentice who has been working there for 18 months. It's a good day, the cherry trees on Chester Road are in bloom, temporary clouds of the palest pink against clear blue skies. I cycle past people taking photos. I find Aimee waiting in the rose garden, spot her all-green apprentice uniform and her slight shyness from the gates. She speaks softly, albeit with a quiet ferocity. Aimee's the kind of person who's unafraid to use your name in conversation and speaks plainly, with neither self-deprecation or aggrandisement. She laughs when she explains she's 'pushing 40' – I tell her she looks about 23. Two dreadlocks snake out beneath her bun.

She takes me on a tour of how she sees the park, pointing out the Nannies' Lawn, where the women employed to take care of richer people's children would take their charges; the fossilised trees; the beds and borders she's worked on. She's one of only a dozen park apprentices in the city; it's one of the rarest – and probably most over-looked – gigs in town. While working here Aimee's become familiar with the park's customers, so to speak – the people who use this space like others would corner shops or train stations. 'I see the same faces pretty much every day at the same time, which is quite nice,' she explains, as we walk the broad paths. The same person chants on the Japanese Garden's viewing platform, early in the morning, every day. There are joggers and dog walkers, a colleague named Phil, who has earned the moniker 'Pigeon Man' from Aimee for his habit of naming and feeding certain birds. They land on his body, Aimee explains, his head and arms. 'It's a sight to see,' she says.

Over the past year or so, Aimee has worked in Regent's Park, in the blaze of heat and out of it. Her favourite month is May, when things are growing. 'If it could be May all year round, I'd be very happy,' she tells me, during this cold April. While working, Aimee often receives comments of gratitude from the passers-by, and others of wistful longing for a job that looks more bucolic than it often is. Sometimes, less often, people ask her where she is from. Aimee is from Norfolk, but as a woman of colour she's suffered the question throughout her life.

Aimee began to entertain a career as a gardener after taking a membership service job with the Royal Horticultural Society, manning the phone lines for members. Although it became permanent, it was a temp job at first; she'd moved from Northampton to London and signed up with an agency. It was during a community gardening day, organised by the society, at the Angell Town Estate that everything 'clicked' for her. 'I realised I liked it, there were good people to talk to, it was very stress-free.' Gradually, she learned more about the apprenticeship scheme through her own research; it wasn't immediately obvious how one became a gardener. 'I remember seeing a gardener in St James's Park,' Aimee tells me. 'I was like, "Oh, I wonder how you do this job?" I wanted to ask him, but I was too shy, so I didn't.' She later learned it could be as simple as walking into an office with little or no experience and asking for a job, she tells me. She's seen that happen with one or two of her non-apprentice colleagues. Beyond the practicalities, Aimee didn't initially feel that the scheme was 'for someone like me, but for those from more privileged backgrounds with trust funds'. She took the plunge after seeing 'some representation' on the Royal Parks website. 'So I was like, "OK! Go for it!"' Though she notes that she was fortunate to have a partner to share rental costs with.

Horticulture, she tells me, was part of her childhood. Aimee grew up under big skies, on a pig farm in Norfolk. Her dad worked in the sheds, and before she and her brother were born, her mum worked in a local plant nursery. 'We'd get free muck for the vegetables; Mum and Dad always grew vegetables and fruit in the garden.' As a child, she loved it; as a teenager, not so much. She learned that her childhood affinity was 'really true' in her mid-twenties: 'I wanted to grow my own food.' There was an allotment in Northampton, and a broken ankle swiftly after, which scuppered things. Aimee moved to London, where she'd vowed never to return after being priced out a few years earlier. But, she says, it's 'been very good to me the second time around'. The city is still a struggle at times: Aimee's frustrated that, in her late thirties, she and her boyfriend have to share a first-floor flat with another couple. She swapped the salary of an office job for that of an apprentice. The goal, she tells me, is to move to a house with some land. 'Get some animals, grow our own food. That's what I'm clinging on to.'

I suggest her parents must be proud, that after taking jobs in offices all over the country, she apes the life she was raised with. 'Yeah,' she half-agrees, with a small laugh. 'I kind of see it as a natural way to live as a human, but I'm biased.' They were surprised when Aimee told them she was doing an RHS course before her apprenticeship: 'I'm interested in gardening now,' she explained. 'I want to grow my own food.' If anything, Aimee traces her connection to the ground further back, to an estranged grandmother. 'I've only recently learned – maybe in the past six to eight months – that she is a really keen gardener,' she explains. 'I've never known her, though. This is a really weird series of events. Do you want me to go into detail with that?'

Aimee contacted a genealogist who specialised in Bajan ancestry and gave what details she knew of her maternal grandmother, only for the genealogist to reply, with some surprise, that they shared a relative. 'I know Barbados is a small island,' Aimee quips, 'but that's ridiculous.' The genealogist spoke to Aimee's grandmother. 'She's told me about how she is a keen gardener. And then I do some more research and in Barbados gardening is, like, a big thing.' When Aimee spoke to her mother about it, she was also taken aback. 'I think she was very surprised. Very, very surprised.' Aimee's mother moved from London to Norwich to study horticulture, after getting into gardening in her late teens. 'Would it have happened if [my parents] hadn't gardened so much? I think it probably would've,' Aimee says. 'Maybe it would have always happened.' Her longing for the ground is part of what she calls an 'ancestral memory'. So much, she believes, is passed down – often without us realising. 'The amazing nature of genetic memory fascinates me.'

•

The hazel leaves are luminously new when I return to the woods beyond Treehouse one Sunday morning. I am there – quite early, the woods are still quiet – to meet Bethany, an artist who responded to the survey with a story about her grandfather. He'd lived in central Belfast, in a tiny house with a small backyard 'lined with wonky shelves filled with yoghurt pots and cans, growing plants and flowers', she wrote. 'He managed to create so much from what felt like nothing.' I've not been to the woods for a few months. Together, Bethany and I retrace the looping paths I got to know when I lived nearby. She tells me about her work, which looks at the space mining industry, about her forthcoming Venice Biennale show. Once we've found a bench to sit on, she gently pulls a couple of St Brigid's crosses from her handbag. Geometric and brittle, these square crosses take the same shape as the hundreds her grandfather would make through the winter months from wild rushes that grow in Northern Ireland's boglands, often brought to him by more rural friends in bundles. He died four years ago, she tells me, on the Feast of St Brigid. 'I don't even know why he made them, to be honest,' Bethany says. 'It's just a memory that happened, and such a weird coincidence that he died on the feast day. There was all the stuff in the home, all the rushes, all the crosses made up. It's bizarre, it's just so special.'

I love listening to these stories, the quirks of people that are remembered when they are gone. The cleverness of growing a rose bush from a stem plucked from a supermarket bunch; the notepads and photographs and bird's nests found in a study. In the week after my grandfather died, I discovered a pencil tin in his long-neglected workshop. Inside were 11 plastic vials, no longer than my little finger, stoppered with cork. Some were numbered, some just white labels faded by time, but one still had its writing: 'BETA VULGARIS, LARGE YELLOW GLOBE'. Beetroot. I sometimes think about sowing the seeds inside them, but it's been six years and I've still not. I think I fear the disappointment of them not germinating; of trying to usher something from the soil that was gone long ago.

As the garden fills out with summer, I make small reclamations of the people I thought led me to it. I serve up salads in bowls that belonged to my grandma, place them upon the bumpy creases left in her tablecloths. My disappointment in my first year of growing sweet peas from seed – too much rain, not enough food, stubby little stems and all over within three weeks – is amplified by a sense that in cultivating them I am continuing a habit started by my grandad decades before. I find a pair of my grandpa's overalls in the back of the wardrobe and take them on the Tube, to a seamstress in North London who will unpick every seam and cut the fabric to make them fit me, instead. I wear them deep into the seasons, into autumn rain and over layers in winter chill. When I wear them, I think of him, but I also think of my gran, and of her botanical brain. My mum is woven among all this, too. Since I went back to the garden I was raised in, it's been easier to notice what she taught me: how to feed a crowd without getting in a flap; the value of a good, second-hand patio container; knowing that if you want to paint a kitchen pink, you should just do it, and be happy. These may sound like small domesticities, but they creep into and around the home I have made here. It was once so easy to dismiss these lessons, ones instilled so deeply that it took me years to recognise them, but they have made me. They taught me how to create a space for myself, one of the most quietly feminist acts there is.

10

CANARY WHARF

Months after I'd visited the soft sprawl of Sui's garden, her words would come back to me. She had pointed out a part of the woodland where an established bamboo reached up against a gnarly, handsome oak tree. 'I like those two,' she told me, 'because I feel like they represent me: a Chinese bamboo, an English oak, coming together.'

I'd been thinking about inheritance in the garden; what we bury, what we uncover, what we seek of ourselves in the ground, but identity is also grown there. The two are siblings but nevertheless distinct. If inheritance is what we are born with, what we carry, wittingly or otherwise, identity is something we construct ourselves. Perhaps I had inherited a love of gardening – from a grandmother whose girlhood love of botany was never nurtured, from grandfathers who grew flowers for the women they loved, from a mother who wanted to add flavour and beauty to her children's upbringing – but increasingly I felt it was something I was making my own.

The garden is beginning to surprise me. I'd spent so long thinking about what it should be that I hadn't anticipated the reality. A year after we get the keys, we witness an eruption. The patches of bare earth I'd found so dispiriting vanish into secretive, shadowy spaces I won't see again for months. It is both unknown and familiar at the same time. This, this torrent of growth and green and life, is all my making. These beds had been scrubby lawn a year ago; now they are filled with wonder. In all the hours I had spent staring and envisaging, I could never have pictured something so beautiful.

Every morning, Matt shakes himself out of bed to put the kettle on. Coffee for him, tea for me. Once he's brought it, I go outside to look at the garden, and when I come back mine has cooled down enough to drink. Whatever control I had over this space, with the shedbuilding and measuring, has gone. I shrugged it off, as I might a coat upon walking through the door. The garden is swollen with growth and I have come to look at it deeply. One milky soft morning I discover a square foot where the foliage of grass, fennel, dahlia, angelica and peony collide. This is where I see a flicker of reflection and of recognition. The movement, the muddle and the small miracle of it all. A few weeks later I overhear Matt describe the garden as 'a wild, verdant jumble' on the phone. I wonder what it is to him, this unfurling space. I wonder if it is surprising, I wonder if he understands that this is my making, my doing; that this wildness is the work of his future wife. He is unimpressed by the borage, which sprawls inelegantly over the border edge. 'All that came from one seed,' I tell him, and he's astonished. 'It was always going to be a kind of chaos,' I add, knowing it in my bones.

•

I didn't recognise the writing on the package. Brown paper gave way to white tissue and twine, and beneath that, a hardback of faded linen, embossed with art deco linework in grubby white. It was from a cousin on my father's side I'd not spoken to in years; she was of his generation, and there hadn't been a family get-together in a while. The book was gorgeous: *Plant Form & Design* by W. Midgley and A.E.V. Lilley. A rudimental bookplate, made of brown paper, explained that Dorothy Pring had won it as a Prize for Drawing in 1902. This was somewhere between an Edwardian coffee table tome and a textbook, intended to point out how to translate naturally occurring beauty into human-made ornament. It held line drawings of seaweeds and poppies, cross-sections of dandelions. Between the pages dedicated to bluebells lay a recently pressed specimen, its petals bright against the distantly smoky paper. My cousin's work, I suspected.

Diane had sent it because she'd heard me talking about the garden on the radio, and because, as other members of the family told me later, she was good at that – giving unusual things to people that she thought, often correctly, they might like. Before I'd had a chance to write back thanking her, there was another unlikely arrival. A forwarded email from the cousin's sister Pauline, who had been living in the neighbourhood for the past six years without me knowing. I called her while watering in the garden, and we spoke about what her life was like. Pauline was 80 now; when I'd last seen her I'd been a child, and yet I must have walked past her house dozens of times since moving to London. She invited me to visit, and so a few weeks later I go with my sister in tow, having carried a handful of peonies and nigella across the South London hill that separates us. When she opens the door, the first thing I notice are Pauline's cheekbones – high, and narrow, like mine, like my father's, like my great-grandfather's. A strange notion of familiarity in a face I wouldn't recognise had I passed it on the street.

We sit in the sunshine-yellow basement of what she calls a 'doll's house' and catch up on family news over strawberries and apple juice poured from a glass bottle. The flowers catch their breath in the kitchen sink. We walk through the garden, admire the poppies that are sprawling their petals, look at the hole where the fox gets in. Pauline's is a house full of beauty and colour and curio: she is an artist, and complains that she is running out of room to hang a lifetime's worth of painting. There are stripes and weavings and terracotta jugs. The place generates a complex kind of warmth.

My family tree is large, but our branches are disparate. Four brothers were born before and during the First World War, and from them amassed a web of children and cousins, sliding with the generations and geography. Some I keep in touch with, others I've barely met. Story and upbringing and loss stewed together over the years and left more unsaid than known.

How funny to have one's family presented to you on a wholly normal afternoon in May. How nothingy, and how huge. A letter, a book, a forwarded email. I called Pauline before I thought about what I might say. I'd been looking for threads of connection and here they were, a mere mile between us. I'd cycled past, walked past, Claverton Street without knowing my ancestors had been born there. As a child I'd missed my connections to London, which was a place we visited on the train, filled with red buses, somewhere that was home to Mr Benn, not my grandfather, who lived in a red-brick house in Reading. Over a decade of living here and I still can't work out if I'm part of this city, even though I have given it my girlhood, my sunrise mornings and night-bus twilights, my love stories and my heartbreak. When I walk to Pauline's house, I walk through the park I looked over in the flat I lived in with an old boyfriend, the same park I would cross after secretive sleepovers at Matt's house early in our relationship, the same park I take to now when I am listless or needing to see the skyline, the same park I might walk in, still, in years to come.

To live in London is to make it small and familiar. But it is also to learn new things. Together, Matt and I moved to somewhere I found – I find – hard to love. A corner of London that felt strange and loud, surrounded by the thick and humming arteries that would whoosh us into the city. Walking distance from the bridges, from the river, and yet without a sprawling piece of parkland in easy distance. I knew I would miss the greenery I had become accustomed to just miles across the map. In Treehouse we were flanked: woodland, gardens, parks and hills. It was far-flung but that expanse held us well. I came to trace its breadth over weeks of daily walks, meandering through the quiet streets of big houses and blossoming trees. The empty evenings would stretch out and we would stretch with them. As Matt and I trod new ground together I bumped into my younger selves: the girl whose first place in the city was a flat on the corner of a park in Peckham; the older one who took her reluctant lover on reluctant bike rides up relentless hills. Matt and I formed a routine consisting of fishmonger and bakery, of the rattle down to both and the climb back up. Once home, we would be welcomed by the trees, whose birds I watched like a soap drama – when they fed, when they sang, when they roosted against the dusk.

The Camberwell-Brixton-Stockwell hinterland held none of this for me. The only familiarity was the nightclub on the corner, which I had been to just once on a loud and melancholy night. I couldn't equate that girl, standing in the queue mere metres from my front door four years ago, with the person who parked up outside it and unpacked pot plants from the car. Summer stultified and became a glowing autumn, then a soggy one. Over that hard winter we struggled for places to go. Caught between boredom and claustrophobia I urged us out when we could muster. Bitterly brisk bike rides away to the suburbs; crisp early Sunday strolls to Buckingham Palace. One grizzly afternoon I put on a thick rubbery raincoat and took myself out in a storm simply to feel the outside on my skin. Mostly, I took to making desultory laps of the local park – a sweet Victorian endeavour with a bandstand and now a basketball court. I watched the ginkgo tree yellow and drop, paint the hard ground beneath it gold. I watched the halo of mist gather around dogs running on cold mornings. I'd take my malaise out, attempt to shake it off before another winter's day spent indoors. I'd coax my weariness off on the bike.

Sometimes I had specific places in mind but more often

I ambled on two wheels, pushing myself into a curiosity that tricked my brain away from the exertion. I would stop at traffic lights and wonder why my chest was heaving. These streets didn't seem extraordinary enough to make destinations of, but they were new and to be uncovered. I pushed on, turning into dead-end mews and making last-minute swerves into new streets. I made decisions on impulse, wending my way in concentric circles. My hand didn't worry for the map in my pocket; I knew I wasn't far from home.

Instead, I looked at the houses and their gardens. A fulsome camellia hedge, a tree caught between pink and green; a smart round bay lollipop and a centuries-old bromeliad that eclipsed the 1930s house behind it. Troughs of drunken hyacinths and daffs going over, tulips in lipstick colours and neglected front yards a riot of forget-me-not and dandelion. Behind them, the houses: the tall and the grand, the proud and the sub-let. The anonymous sheen of newbuilds, the storied curlicue of Victorian plasterwork. London is a city of streets bombed, built over, torn down and extended. Immigration and gentrification collide side by side. Smart new neighbours and scruffier older ones. These are the things you can read in the state of window-sills, in what pushes through the gravel.

As I rode, I realised I was making a new map. Some parts of London are etched on my brain in memories, in the clothes that I wore there or the arguments I had. The people who I stood a few feet from, bent double and crying with laughter, eyes opening to the pavement, upside-down with happiness. There are the miles worn in by commutes, the road names I can only hear in the tone of the bus announcement, the rows of houses in which once, some time ago, a party was held where I would meet the man I would marry. My mental geography piled up like onion skin.

This new ground showed me things. The house with the rose-printed curtains and a wall where camellia flowers were picked and laid out as if on display. The pink winter cherry on the corner of a road. I would often lap Myatt's and think of the woman I might become there: would I take a child to the nursery in its centre? Would I push one around in a buggy while they were small, and I was blindsided by the prospect of taking this new, loved being out in the world? Within a matter of square miles my life had unfolded in plot twists. When I walked to Pauline's, I would also walk past that first balcony where I taught myself to garden, which held me when I broke. Five years on and I am still unable to bring myself to look up and see what it looks like now.

It is that balcony I am reminded of when I go to Fernanda's flat. Fernanda's a little shorter than me, although we share the same quietly efficient energy. A few days ago, she tells me, she quit her job in market research. When she did work there, she would turn the camera off during

Zoom meetings and go out on the balcony. She lives 29 floors up in one of the searing, anonymous tower blocks that comprise the oddity of Canary Wharf. Steel and glass glint off the dull metal of the Thames that cradles it, greenery confined to lawns behind marble walls and lollipop trees. The door to Fernanda's flat is at the end of a white corridor, a short walk from a lift taken in silence from a glossy reception, but it opens up to life: her balcony brims with pots, with green, with things growing in spite of the altitude. Leaves bob around in the wind, mostly at ankle and knee height: the ruffle of kale, the stretch of chard, a mop of coriander. A tomato plant holds to its trusses against the wind. In the corner that faces the river a clutch of sweet peas is flowering – Fernanda's first, she says, excitedly. There's the 'impulse purchase' rhubarb she found in Waitrose. 'The security guard felt very compelled to give me a lot of advice, like fertilise the leaves,' she laughs, her shifting, ocean-spanning accent lilting upwards. 'People who like to garden love to give other people gardening advice.'

The pots are mostly plastic, few things are labelled: this is the kind of happy chaos I relish in the very best gardens, where the desire and curiosity simply to grow things outweighs that for aesthetics, order or rules. Beyond and above, the city: the arching spire of Crystal Palace to the left, the tower and office block peaks of East London to the right; immediately in front of us, the architect's drawing of Canary Wharf. This balcony is a green dot on a grey landscape.

Even now, even after a decade in the city, I find myself dazzled by the view. But for Fernanda it's a kind of home-coming: in Hong Kong her family live on the 25th floor. 'I grew up with all this, like, skyscrapers, so this,' she says, gesturing at the plants, 'is what brings me joy.'

Fernanda's only 25, but as an adult she has lived in many different cities, moving once a year for the past five, from America to Denmark and now England. She has only been here, tending to this balcony, for four months. Amid this upheaval, Fernanda taught herself to garden, killing houseplants with care in American dorm rooms; creating orderly, neat container gardens in tiny plots in Copenhagen and experimental compost systems in Old Street, where our gaze rests. Each new home has given her new opportunities and challenges. 'I never had to deal with the wind until I moved here,' she says, throwing me a glance.

Gardening in Europe has given Fernanda a new lens on cultural difference and identity. When she started working here, she was surprised to see the same plants sprucing up her office that grew easily in the lush subtropics of Hong Kong. Sweet peas, by contrast, were 'so special and so new'. She tells me the different toler-ances to mess she's observed between English, American and Danish gardening, and that she's reflected those depending on where she's lived. 'One thing that makes gardening important to me is learning about where species *actually* come from, because like, peonies and roses are actually from China, right?'

Fernanda says she finds it interesting that often, when a person of colour is featured on *Gardeners' World*, their garden reflects their heritage. 'Your garden, for the most part, is a representation of what you want closest to you. It's the breakfast of nature, because it's the first thing you see, or maybe the last thing you do in the evening. And while I really love foreignness in other places, I think in the garden, for me, I'm happier with plants that I recognise, that feel familiar.'

Fernanda was at the end of her adolescence and living in America when she discovered she wanted to garden through two galvanising encounters. 'I took a biology class being like, "Oh, I need to finish this requirement." It changed my life.' Her teacher, she recalls warmly, 'was the weirdest, but she was lovely'. She taught Fernanda about 'invasive' plants and the value of new species, she explained the vulnerability of biodiversity, she sent her on field trips and tested her on plant identification by 'literally holding up twigs in an exam'. She made Fernanda see plants differently: as sex organs, as survivors, as endangered species. The other catalyst happened at a farmers' market, where she tasted a 'good, sun-ripened heirloom tomato' for the first time. 'It tasted like summer,' she says. 'Ever since then,

I guess I grow tomatoes to chase that memory. Food, gardening, it connects you to such a strong sense of desire of wanting something you can't get because it's not grown in this place, or it's too far away.' Fernanda explains Hong Kong's 'huge food insecurity issue' as somewhere reliant on imports to sustain its population. As a result, the understanding of where food comes from is weaker. 'Sometimes when I share videos and photos of my balcony to my friends, they genuinely don't know where their food comes from or what it looks like.' It means that Fernanda has spent her young adulthood seeing food and plants anew. 'It sounds really cheesy, but for me, every single time I eat something I'm kind of brought back to that moment of when I first tried something.' Fernanda's balcony, then, holds both things: a quest for the home-grown produce she's enjoyed in the West, and a determined desire for the flavours of home she can't find easily in British supermarkets. After three years of trying, this spring Fernanda's shiso – a herb in the mint family that grows in the mountains of China – has come to fruition. When I ask her why she's persisted, her reply is simple: 'It's really hard to buy here.'

But Fernanda also gardens for experimentation: she will be moving again at the end of summer to Scotland, where she will start over with a new plot and less sunshine than here. What we are standing among is the outcome of seed trays and patience; the coriander, she tells me, started out as a seed in a jar in her spice drawer. 'There's a part of me that always tries to be pragmatic, you know: garden for your climate, garden for your space. And then there's a part of me, it's like, "Oh, what the heck, let's do it anyway." Because why not?' she asks. 'What I love about gardening is failure is almost never made total.' Fernanda points to a rather sad straggle in a pot. 'This is failing. But I have some bok choy growing over there that's full-size.'

It isn't easy to grow food in a small space that doesn't have direct access to soil (the sheer quantity of bagged compost that was arriving at Fernanda's building caused the concierge to ask what she was up to), let alone 29 storeys up in a rain-shadow. But I don't get the impression that this was much of a consideration to Fernanda when she started to grow things. She simply wanted to: to try and capture that taste of a sun-ripened tomato, to bring sweet peas into flower, to see if she could grow what she couldn't buy. We talk about houseplants, about the stereotype of who keeps them, about burn-out. This is probably the greatest commitment to growing for the sheer thrill of it that I've seen. Fernanda knows she will have to move on before the season is even out; her gardens have always existed despite her state of flux.

I put on my shoes and switch off the dictaphone. We share a few words as I leave, Fernanda leaning on the edge of the open door as I stand in the hallway. She speaks of soil, and residency. Later, I email her for clarification. 'I find it hard to talk about soil without talking about an underlying question of belonging,' she writes back. 'There's a big emotional and political question still of how it is possible that BIPOC "own" (in inverted commas because ownership is a particular way of relating to land and possessions) so little land/wealth created from land, even though their ancestors farmed or tended to the land as immigrants, indigenous people or as enslaved people.' Fernanda references a few things to further her point – the notion of white escapism inherent to the romantic rural life fetishised by the cottagecore trend on social media; a scene from Bernardine Evaristo's *Girl, Woman, Other*; organisations that reclaim access to land. At the door, she says, she was talking about 'how historically land/ soil have been marshalled in the discourse around citizenship and belonging'. It's an academic email, where our conversation had been shifting and often excitable. In the gardens Fernanda made – those on a desk in an American dorm room, or a yard in Copenhagen, or on the side of a tower block in London's docklands – came space to hold and push against these histories and contradictions. I saw what she'd grown on her balcony as rebellion and perseverance. Those vegetables and flowers were as defiant as they were beautiful. I saw a woman staking claim on whatever patch of land she could access.

Fernanda was among the women who had replied to the survey I'd posted a few months before. In her response, she wrote that after moving to a new place, getting a plant 'helped me metaphorically and literally set down roots'. It also, she wrote, helped her to mark that part of her adulthood: looking back on photographs of her different urban gardens 'reminds me of that time in my life'.

I'd gone to speak to Fernanda because I was interested in how plants and gardens rooted us, in new cities, in our families, in different places across the world. What I found was a woman who grew a garden without knowing what it would turn into. It wasn't so much that the plants had rooted Fernanda but that in growing them she created a space of her own – of experimentation, of food, of beauty, of delight – in new surroundings. I found so much courage in that: to invest in pots of soil regardless of what the future might be. Fernanda's identity didn't define what she had grown but her experiences had. The balcony was a reflection of the life she had lived and the desires she had nurtured. No wonder it felt so beloved, so deeply wild, among these corridors.

•

At times, the garden is like a mirror. It shows me things that I am surprised to recognise, like standing on Pauline's doorstep and seeing her face with a deep, unexpected resonance. Somehow, beneath the beat-filled skies of Brixton, I have grown a kind of cottage garden, where hollyhocks bump against brick walls and peonies clash with poppies. Fennel wafts and softens through the patch, even though I know it will cause me future problems with its sprawl and seeds. There is nothing sharp or contemporary or 'urban' about this plot.

I think back to Fernanda's comment, that we are happier with plants that we recognise. I ran away from the gardens I was raised in, but parts of them came to claim me anyway. Still, I have carved out this space – these flowerbeds, these pots, these scrambling things along the wall – above and beyond my inheritance. There is much I can't untangle. The deep scarlet of the martagon lilies that brood beneath the tree; a flower I'd never entertain in the house but hungrily await in the garden. The lush ferns that I am endlessly drawn to, that gradually take the place of blowsier flowers; they remind me of dancing in festival woodlands in my twenties, of the earthy promise of May as a student. Increasingly, I see the women I've met in the plants in my garden – the cuttings that have been shared, the plants I've adored and have later bought myself. With time, I will see these plants and tell stories of my own.

11

CHARLESTON

THE SWEET, CLOYING SMELL OF cow manure always hoicks me back to my childhood. I catch it in the air at Charleston, the tawny 16th-century farmhouse in the Sussex Downs that, for most of the 20th century, was home and playground to members of the Bloomsbury Group. When it's open – which it isn't today – Charleston lures visitors from all over to see its painted walls and gorgeous textiles, the rustic ceramics in the kitchen. Here are the bedrooms of economist John Maynard Keynes and artist Duncan Grant, here is where they sat with critics Roger Fry and Clive Bell, here is where they swam in the pond and made work that changed the shape of the 20th century. Charleston is the kind of place that sends you packing determined to paint circles on your doors in mustards and dusky pinks. I first came here as a precocious teenager, keen to research an A-level project. Fifteen years later and I've returned, following a similar kind of curiosity.

Then, I was lured by the bold colours and brushstrokes. I romanticised the bohemianism of it all. My understanding of the Bloomsbury Group's politics and sexual entanglements – and how the two informed one another – was naive, but I loved Grant's paintings and attempted my own imitations, making my friends lounge around on their parents' suburban armchairs in ballgowns, painting them in oils of russet and ochre. I was taken in by the wary stare and clasped hands in Grant's 1918 portrait of Vanessa Bell; the way the straps of her red dress balanced on her shoulders. And it was Bell who had brought me back. In the decade or so since, she had been reclaimed by an art history that had overlooked her as Grant's life partner, as critic Clive Bell's wife, as the mother of their children, as the lover of critic Roger Fry and the big sister of Virginia Woolf, and instead realised she was a pioneer of abstract art in Britain. The first major retrospective of her work opened in 2017, almost 60 years after her death. Finally, Bell was being taken seriously as an artist, her desire to blend her domestic and maternal life with her practice being reinterpreted as a revolutionary – still – way of living. This was undeniable progress, but I wanted to wade into less-chartered waters. I wanted to learn about her as a gardener.

Charleston is surrounded by its grounds: a gulp of a lake, still carrying a tethered rowing boat on its water, an orchard of gnarled medlars, a snicket of tall trees. Sculptures made by Quentin Bell, Vanessa's son, loom between the boughs. From most places, you can see the Sussex Downs rumple out beneath like a quilt. The best bit is the walled garden, which fans out from the back of the house. Piebald walls of flint and brick hold roses and apple trees and cloud-pruned box hedges. There's a neat square of lawn with a tiled rectangular pond in the middle, but even now, on one of April's last warm days, the overwhelming sense is one of gorgeous rebellion. Bell and Grant's daughter, Angelica Garnett, wrote that Charleston's garden 'was not a gentleman's garden or a gardener's garden, it was always an artist's garden'. Women make gardens for many reasons, but here was one that appeared to be a muse.

I walk through the gap in the walls that surround the house and spot Harry, Charleston's gardener. He's bent down but when he stands up his head of unruly curls is level with the top of the garden wall gates. The tulips are out, the last of the narcissi, the apple trees are budding, but no blossom yet. Winter–spring under clear skies. It's been dry: no rain, no April showers nearly all month. A sprinkler chugs guiltily against the birdsong. I call out to Harry and he waves a long arm; he's agreed to speak to me about Charleston's garden and Vanessa's role in it. We sit down on a stone bench in the garden's sunniest corner, where he's laid out papers and books with gardener's hands. Here, they are tools as vital as a wheelbarrow or spade: when restoration work began at Charleston in the eighties after Duncan Grant's death, the garden was a mess.

Navigating back to the garden created by Bell and Grant to Roger Fry's 1918 design has been a decades-long archaeological dig of its own through paintings, letters and lost seed packets. Harry inherited this challenge relatively recently: his predecessor was Fiona Dennis, a plantswoman devoted to tracking down the varieties Grant and Bell had planted nearly a century before.

On the bench, in the early afternoon sunshine, we twist to look back at the house at the top of the garden. Harry points to the windows where the main studio was; the attic where Bell worked in her later years; the ground-floor room where she slept. He speaks so softly I'm not sure the microphone can hear. 'In many ways she was probably the one that woke up to the garden every day,' he says. 'I think in that respect, the garden must have had a real, sort of tangible, significance for her. Perhaps more so than for Duncan and the rest of them.' It's a funny thing to hear, resonant somehow. I have come to Charleston on an inkling – more instinctive than proven – that Bell's relationship to the garden was something deep and complex, something possibly overlooked, as her art has been for so long, because of the work done by others around her in the home that she helped to create. Bell died in 1961, at Charleston. She wrote letters, she painted prolifically; her granddaughter, Virginia Nicholson, is Charleston's president and holds childhood memories of a grandmother who would encourage her to sit for portraits by asking her to tell stories of the paintings on the studio walls. But that is all there is to work with: memories and makings of a woman who died 60 years – plus a couple of weeks – before my return to the garden. I couldn't ask her about her garden as I had with other women. Instead, I would have to piece together document and history, sense and logic, to create something that felt true, much as Harry, Fiona and the gardeners before them had done.

'I wish you'd leave Wissett, and take Charleston,' Woolf wrote to Bell in May 1916. 'It has a charming garden, with a pond, and fruit trees, and vegetables, all now rather run wild, but you could make it lovely.' By the autumn, Bell had done so, along with her husband Clive, Duncan Grant and his lover David 'Bunny' Garnett. The walled garden was a quagmire of mud and potatoes, the apple trees – included, to Bell's delight, in the rental agreement – were fruiting, but it was nevertheless 'a desolate place', Quentin Bell remembered. His mother had packed artichoke roots, the beginnings of the towering cardoons that would punctuate the garden and her paintings for decades to come. The Bells had two boys: Julian, then eight, and Quentin, six. The group had been lured to Sussex by the raging war: Grant and Garnett sought agricultural work to enable their legitimacy as conscientious objectors. For Bell, a born and raised Londoner, growing food to feed her boys was the priority. 'Life in wartime Sussex seemed to Vanessa principally a matter of survival,' remembered Virginia Nicholson, in conversation with her father Quentin Bell in their book, *Charleston: A Bloomsbury House & Garden*. In a letter to Roger Fry in late February 1917, Bell writes: 'I'm making great efforts to get hold of a pig, rabbits etc., and get the garden dug. I see the only way is to take complete charge myself.' A determination that makes me imagine frost on mud, grubby knees and short days, and the ready makings of a matriarch who would keep the house ticking over for a further four decades.

Not long after, Fry started work on a more formal ornamental design for the walled garden, and began to lay the brickwork of the grid-like hard landscaping that remains today. Charleston's garden, like its house, was a collaboration. Beyond Fry's design, Grant was a magpie for plants, seeking out tropical varieties that recalled his upbringing in Myanmar. He cast heads and figures from art school plaster casts and dotted them around the place. He longed for flamingos that never arrived. There was one gardener, and then another, to do the labour and maintenance. Bell's role seems both less defined and somehow more centrifugal: she oversaw the gardeners; she pruned and weeded; she cut flowers to relish in the house; she painted, she painted, she painted. 'Charleston is as usual,' Woolf wrote in 1922. 'Nessa emerges from a great variegated quilt of asters and artichokes.' Angelica, Bell's daughter, recalled her mother in the garden 'hovering peaceably in front of her easel, her dress protected by a flimsy French apron, her feet in flat-heeled espadrilles, and on her head a broad-brimmed hat to shade her eyes from the glare. Her presence was betrayed by a smell of oil and turpentine.'

There are dozens of flowers in Bell and Grant's paintings, but the garden lives as much in her writing. 'Bell's responsibility for the production of such fulsome flowering and fruiting is equally recorded in her letters,' wrote Charles W. Moore, William J. Mitchell and William Turnbull in *The Poetics of Gardens*. The garden, and her love for it, is strewn through Bell's correspondence. She wrote about buying lilacs and planting bulbs, kept friends informed of the place she said she couldn't bear to leave. 'The garden is an overwhelming blaze of colour,' she wrote to Julian, who in 1936 was living in China, 'pinks out in masses, roses'; in another, she tells her son the garden is 'simply a dithering blaze of flowers'. Her letters would contain lists: 'the medley of apples, hollyhocks, plums, zinnias, dahlias, all mixed up together'. Harry reads out his favourite snippets of these letters to me. There is one from August 1930 that sounds like a colour chart: 'The garden is incredibly beautiful . . . it's full of reds of all kinds, scabious & hollyhocks & mallows & every kind of red from red lead to black. Pokers are coming out . . . I have of course begun by painting some flowers. It seems the inevitable way to begin here.'

It's a planting choice that is key to Charleston's garden being one made by and for artists – flowers look brighter against grey foliage than they do green – and one wildly out of step with horticultural convention at the time. These were Mediterranean plants, ornamental varieties muddled up with fruit trees and vegetables. It was rogue, even by the standards of a century later. In the farmhouse, the corridors and woodwork are painted the same dull grey – a colour that has earned the house's moniker in shade cards. The spirit of the house, the attitudes of those who lived there, drifted out into the garden. The inside and outside are merged. In this domestic sphere all was under- taken with the same desire: to find a freer way of living.

The freedom of Charleston – of those who lived and loved and worked there – is well-documented. There are photographs of parties and plays, of children swimming in the lake. Before he died in the nineties, Quentin recalled the various dangerous and morally dubious antics he and Julian got up to as children, running amok in an orchard that had grown into wilderness. But Charleston was also chosen for its seclusion. Nestled in the Sussex Downs, it offered as much of a retreat as it did a rumpus. Friends often visited but it seems like Bell had boundaries; she infamously pretended to be in London when Pablo Picasso, staying nearby at Lee Miller's farmhouse, wanted to drop by. Bell was known as 'Nessa' to her friends and family, and winkingly nicknamed 'The Saint' by her siblings; art history has remembered her as some kind of domestic earth mother, wafting around the garden in her trademark hat – an image that Woolf was also guilty of ascribing to. But, as it is for many women, Bell's maternity was complicated. 'I am honestly terrified sometimes of the responsibility of having any children,' she wrote to Woolf. 'I'm not sure it doesn't mean hanging the most terrific millstone round one's neck.' Nessa's portrayal as an effortless matriarch was also punctured by her daughter. In 1984, Angelica published *Deceived with Kindness,* an incendiary memoir of her upbringing in Charleston's clutches that suggests Bell's life was chaotic and messy, her parenting so problematic it left the younger woman 'emotionally incapacitated'.

The mid-thirties swept in with a brutal handful of tragedies for Bell. Her friends Lytton Strachey and Roger Fry died in 1932 and 1934. Her eldest child, Julian, was killed, aged 29, in the Spanish Civil War in 1937. In 1941, Woolf died by suicide in the River Ouse beyond her home. Julian's death devastated Bell. 'Vanessa was completely shattered,' Quentin remembered. 'We took her back to Charleston and for the rest of that summer Virginia devoted herself completely to her sister and gradually restored her to a quiet and very sad convalescence. On one of the few days when she could not come over to Charleston, Virginia sent her sister a note. I found Vanessa quietly crying over it in the garden. "Another love letter from Virginia," she smiled very faintly.' This part of Bell's life comes up quickly in my conversation with Harry. 'She had a five-to-ten-year period of just a lot of grief,' he says. 'It was about that time, the late thirties, that they really did start to base themselves here.'

With the forties, Bell's letters changed: no longer addressed to Fry, to Julian, to Woolf. She occasionally wrote of her grief and her recovery with Vita Sackville-West, Woolf's lover and sometime muse. Long the family photographer, Bell stopped taking photographs after Julian's death. And perhaps the garden changed too. No longer the muddy deluge relied upon for vegetable growing, nor the wilderness her children claimed their own; no longer the project conceived by her lovers and created in tandem with them, the emerging rose garden she had always dreamt of and described in letters to Julian. Instead, the garden reflected a new stage of motherhood as Quentin, Bell's second son, began to make changes: there was his vision of it as a 'second Versailles', with an extended lawn that Bell looked out onto from her bedroom; the sculptures, paths and piazza that he put in place. But perhaps the garden also became a site of solace, a retreat within the retreat that Charleston already was. In the wake of Julian's death she planted a row of poplars in his memory, just beyond the walled garden. They never took hold of the land but died one by one. 'It is difficult not to be depressed by so much bad news,' she wrote in a letter in 1950. 'The trees are almost entirely bare now and often very beautiful – do you remember how Julian liked the bare trees? – but all the fruit has been picked and there are hardly any flowers. Still the beauty is very great, more so I think than when all is green.' Often, before she signed off her letters, Bell gave a brief update on the garden: 'now a miracle of beauty' in May 1959; 'I fear the garden will be weedier than ever' ahead of her mastectomy in 1944; 'The garden here is simply lovely, only things come out too quickly and are almost over at once' a few weeks before she died.

In her later years, the house became less of a free-for-all: family still arrived at Charleston's pale pink door – old friends, grandchildren – but fewer young people, fewer high spirits. Bell began to resent the road beyond the house, both the noise of it and the people it brought. In 2020, in the heady silence of a locked-down summer, Harry wrote about his solitary gardening of Charleston: 'I have begun to feel more clearly how the garden can become such a consoling remedy for our sorrows.' Sitting in the garden, Harry tells me that he sees the gardener in Bell in what she's left behind. 'Once you start rifling through you really come to realise how significant the garden was, particularly for her, just based on how much material there is in terms of her paintings and her letters as well. She's always talking about the garden in letters from the period. And I guess it must have been a real solace somehow. That's kind of the way I view it.' Other remnants suggest Bell's comfort in this space. One of the last photos of her was taken in 1960, the year before she died. Bell's sitting on the terrace outside the house, a checked dress grazing her ankles, books and papers tucked under an arm.

Bell set up a studio in the attic, from where she could look over the garden and the pond. She installed French windows in one of the rooms downstairs and turned it into her bedroom. It's here that the musician Patti Smith made her own pilgrimage, keen to capture the 'silvery quality' she saw in the light. And it's here that I stand with Melissa, who gives me her own tour of the house, staring out at the garden. Melissa works at Charleston, and admits that Bell's room gives her 'the heebie-jeebies'. It does feel arrestingly intimate: the bed, which faces out to the doors and the garden beyond, is where Bell died. Her famously oversized hat sits atop the quilt. On the wall, across from her pillow, hang two longing portraits of her boys that she'd painted when they were babies. I reply that it feels an overwhelmingly feminine room; the walls are painted a soft pink. Here, in a house filled with the voices and the exploits of men – of Bell's grown sons, of Duncan and his lovers, of Clive and John and Bunny – was a space a woman had created for herself, where she could wake and walk into the garden that she loved.

Mid-afternoon, Harry walks with me to the gap in the wall before returning to the work I've interrupted.

Charleston's director happens to be standing there and I'm introduced, asked about what brings me to the farmhouse. I fumble in trying to say what I've come here to find: some traces of Bell in the garden, stories of what it meant to her, something to back up a belief that her work here, as her art has been for so long, has been overlooked. The director presses me, asks me how I know this, tells me it's impossible to prove. Grant and Bell did so much work together, how could I separate her work from his? He recommends I look at Roger Fry's archives to learn more about the garden. The exchange is awkward; I feel small and amateur. I have not pretended to be an art historian and yet the implication is that I should act like one. Later, it feels strangely apt that this should be the ending of a day spent trying to unearth a woman's invisible work: with a man suggesting I refer to the receipts of other men's work for validation. So much of what we do – in gardens, in homes, in relationships – exists off the record, in memory and sense. Maybe the only way to find it is on instinct, on the feelings of the air that hangs in a pink-walled room.

•

Place can be important. When I asked women if they would speak to me about their relationship with gardening, I told them I wanted to do so in person, in a green space of their choosing. Sometimes, these would be their gardens, but more often they would be elsewhere: community gardens, public gardens with visitor fees, public gardens without. These spaces, which held memories and meaning, were ones which people felt had formed them somehow. They were integral to their story; they contained something of who they were. Some of the gardens that form us are no longer ours to tend; only one woman refused my request to interview her, and it was partly because she didn't have access to the garden she wanted to show me. 'It is no longer mine,' she wrote, 'and I've no idea how much of me is still in it.'

The conversations I had been having with women in their gardens held many things. Determination and grief, creativity and retreat, loss and joy and control and change all overlapped in the lives I was hearing about, often within the boundaries of the green spaces we were sitting in. These things lurked among the plants and in the dark, complex web of the soil. What brought them together, though, was space. Tend to a garden – create it from the ground up, inherit another's plot, sit back and let it return to the wild – and it becomes a space to enact and to mirror the matter of life. I kept bumping into this desire to carve out and claim a space, in every conversation I was having. I traced it in my own life too. At a time when I felt both claustrophobic and deeply alone, I could find space to air my thoughts and generate connection in the garden. Space continues to crop up in the conversations

I have: a need for it, a desire for it, what we have gleaned and borrowed and offered to these spaces where we grow.

I think about Abi, who I met late in the long winter. She told me about how, between hospital shifts, she painstakingly created a patio garden from the scrubland behind her block of flats. For six weeks, Abi levelled the ground by hand, she collected free paving slabs she found online, she sifted through the excess soil for pebbles to create hardcore for the patio. Gradually, she painted and foraged and planted. By the summer, she had physically made a space that hadn't existed before. In the summer, I meet two women who garden together – Ali and Sarah – and space weaves through our conversation. They tell me about a plot of woodland owned by a pair of friends who couldn't agree on whether it should be managed or not ('a place where you really do belong only to the space that you're in, and you're taken apart from the world where you thought you knew you were, and how are you going to find your way anywhere? And isn't it beautiful?' says Ali, describing what it was to be there). They speak about a guerilla gardening group in Cambridge who tend to spaces left unloved by students and residents. When we talk about the loss of women's gardening stories, Sarah suggests that it is because gardening is often passed down silently. 'You're making space,' she says, 'and women aren't really supposed to do that.'

A season after I first meet Kayla at the glasshouses in

East Sutton, I return. The in-between time has seen a slow patter of communication with prison officials, trying to get access to return to interview Kayla properly. It's early on a bank holiday Monday and the place is deserted; signs for visitors lead me to a locked building. One of the key instructions before I arrived was to leave my phone in my car, but there's nobody around to glean directions from. Back at the farm entrance, a woman in a grey tracksuit directs me to the attic floor of a barn. Inside, tired office furniture is scattered around, the lights are off. I push through a door and find Kayla curled into chair, two men at desks in front of her, telling me they thought I'd got lost. I suppose I had.

I wanted to speak to Kayla again because our last conversation had left a deep impression for all its brevity. I was less interested in what had happened to her and what she had done to land her in the prison system than how she had existed within it. None of the other women I had talked to were so newly acquainted with plants or growing, had encountered gardening in such extreme conditions. None of the other mothers I had spoken to had been separated from their children.

But I'm also conscious that this is probably the most unbalanced conversation I've had so far: Kayla is here, we exchange pleasantries, but I've been liaising with the prison staff, not her – she isn't allowed access to the internet. I don't know how much she knows of my request, of why

I am here. When I first sit down, she asks how many other prisoners I'm talking to. I find myself explaining, a little awkwardly, that she's the only one; that I'm interviewing women who have a relationship with the ground, and hers happens to have been forged in a prison.

She tells me how the business has been growing – the online orders have exploded, somehow, even though Kayla and her colleagues can't manage the orders or reply to emails themselves. Two more women work in the glasshouses now, to help keep up with the packing and plant care. It's not been entirely easy for Kayla to allow them certain responsibilities; she'll be out before too long – 'seven months on Saturday. I'm literally counting down the days now' – and she admits it's going to be difficult to leave her custodianship of the plants behind.

Kayla describes her previous, plant-free existence. 'Wasn't into it at all. Didn't even like having grass in the garden, just wanted AstroTurf and patio,' she says, wide-eyed with the memory of it. 'It was nice to get a bunch of flowers now and then but really that was as far as it went.' Why did she hate it so much, I asked. 'Because to have anything planted was just another thing you've got to take care of,' she replies, a little wearily. 'I think that's how I felt about it before. I've got enough to do, I've got a dog, I've got two kids, I've got a house to run, I don't need anything else that I've got to take care of, and to actually be doing it and not feel that it's a chore, that I actually really like it? It gives you a sense of wellbeing, to know that you're taking care of something really well and to watch it, you know, blossom and bloom.' The biggest surprise of all this, she says, is how much she enjoys it.

The glasshouses also offered Kayla an escape from the confines of the 'house', as she calls the main prison building. It offered her control at a time when nothing was certain. In the glasshouses, there was a quiet consistency that no longer applied at the house: there were temperatures and water levels to check. Among those propagating cuttings, Kayla could find her own kind of release. 'The effects of not seeing your children, you could just worry constantly,' she says. 'Because as soon as you go back to the house, you think, "What's gonna happen with the kids, what's going on with my mum, what's going on with visits?" But I made a conscious effort not to do none of that when I was outside. The garden was the place where I literally could get away from that. And it was great.'

Other women have come and gone, and the greenhouses have cradled those losses, too. 'The one constant has been me being out with the plants,' Kayla says. As her knowledge has grown, she's been adding to her plans for the garden she'll make upon release. 'I don't know where I'll be moving to yet,' she tells me, 'but fingers crossed I'll have a garden.' She wants to grow strawberries and tomatoes with the kids, pick up some of the planters she's become aware of. That garden, too, will offer a necessary space: Kayla will be serving the next four years 'on licence'. 'I'm going to be really nervous,' she says, 'because it's so easy to get recalled back to prison if you're in the wrong area or someone makes a phone call. So I probably will be spending a lot more time at home. And to have that, you know, to spend the day out there, pottering around doing the plants and that? I'm really looking forward to that.'

Before I arrived, I thought Kayla and I might talk about motherhood, an extension of the chat we'd had before. But it's the notion of space that fills our conversation. A glasshouse gave Kayla a timetable to do time by; a garden will calm her anxieties about the next stage of her sentence.

•

I don't often walk into houses like Caroline's. An imposing Georgian rectory at the end of a biscuit-coloured Somerset village, it is a space of falling light – spring sunshine on the floorboards, dust motes dancing lightly on the air. The paint colours and reclaimed fireplaces are the stuff of magazines, but trappings of a no-nonsense family life claim the corners – hoodies in pastel colours, piles of trainers in the hallway. The double-height ceilings jerk my neck up; I find myself pulled towards the Aga. With a frank pragmatism I come to learn is typical of her, Caroline explains as I cross the threshold that she lives here because of a decade of buying and selling houses, and I sense that there will probably only be unashamed honesty between us.

We'd met once before a couple of summers ago, when she'd taken me onto the roof of a nearby country house she worked for. Ever since, I'd felt a kind of warmth, caught between big sisterly and mothering, in our communication: she'd keep asking good-naturedly if Matt and I had got engaged yet, making jokes about dusting off her hat. When I got off the train I found her waiting in the car park − chic in a navy dress and sunglasses − before engulfing me in a massive hug, the words 'Hello, gorgeous!' softening around my ear.

Before this, early on a dark February morning three years earlier, I learned that Caroline had been suddenly widowed. She runs a successful Instagram account, one that charts the seasons and growing things in the countryside nearby, and she'd posted the news on there: he had been killed in a car crash on the way to work, leaving her and three daughters behind. I'd often thought of her and her family since; it was such a horrendously banal method of devastation.

When I contacted her to see if she would speak with me more formally, I thought she might choose the estate she documents every day as a location. Instead, Caroline suggested her own garden, and so we are here, eating lunch at the table on the patio. During the hour or so we spend together, I don't ask Caroline a single question. Instead, I listen. She starts by explaining the house and moves on to a kind of explanation of her life, while fixing cups of tea and tossing salad and fetching cutlery. Her garden is flat and unfussy; a graceful lawn, a big flower bed, tall trees, the top of the church tower over the brick wall that marks the boundary. When she turned up 15 years ago with a six-week-old baby, two older girls and a team of builders to manage for the next year, the garden was 'fiddly'. 'It was given to a team of gardeners, which I knew straightaway wasn't going to work for me,' she says. Since then, she's been paring it back. After sorting the house, Caroline explains, with mirth, 'you start for the rest of your life on the garden, which just carries on and on, and then ends up taking up the rest of your time endlessly'.

In front of us stands a crescent of pillar-like yew trees which, she tells me, she planted in the wake of her father's death. He suffered a major stroke while on holiday in Portugal. 'I got a phone call the next day, telling me to come and say goodbye to him. The stroke should have finished him off, but it actually made him locked in for three years,' Caroline says, the ghosts of outrage still lingering in her voice. For the rest of his life her father lived in a nursing home, 'kept suspended between life and death' by medication. 'I was so angered by the process and so energised by it,' she recalls. 'There's something quite compelling about digging. I don't know what it is. And so I dug these holes and put these trees in, I created a herb garden. I just had so much anger, and it felt good to channel it and direct it and actually make something so it actually then becomes kind of poignant.' She walks me around the garden, pointing out the work done by her boyfriend – 'he's brilliant at digging but just tends to dig everything,' she laughs, affectionately – the fire pit where they cook food on the weekends; a new plot where apple trees will fruit; a veg patch, another that will take annual flower seedlings in a few weeks, when the frost risk has passed. Inside, I'd clocked the seed trays filled with eager-looking cosmos and *Ammi* seedlings under the windows – on worktops and benches. 'All the sort of blossomy, lovely, frothy kind of stuff,' Caroline says, then pauses –'hopefully for our wedding!' She is giddy, briefly, lets out a small laugh. It'll be very quiet, she explains. 'But I am growing all the flowers.'

Caroline's the kind of person who feels 'discombobulated' if she's not been outside for more than a day. 'I've had a garden throughout my life. And it sort of moves with you. It changes with you. You think you're growing it, and it's growing you,' she says. Caroline can chart the 'definitive' moments of her life through what her garden has been. She rattles through her life so far: her twenties, fuelled by 'energy and fire in my belly'; her thirties 'nurturing, growing children. And then you get to your forties, and I don't know, you just feel a lot calmer.'

The garden was there in the aftermath of Caroline's husband's death. 'I can't explain it, but for two weeks I literally stopped seeing things. It kind of felt like you're underwater,' she tells me. 'When something happens like that, literally out of the blue, you go on automatic pilot. Because you have to process the children, you have to process the police, you have to do all the paperwork, you know, at the least appropriate time. You have a lot of things to do, and a lot of things to do right. Because it's so important to other people that you get it right. And so, you get whipped up into this vortex where you have to just do things without processing it, without letting it filter.

'And it's funny. I remember being in the kitchen, actually, two weeks later, looking at the branch of a tree. Somehow, in that moment, I was starting to see things outside. I was starting to see bits of the season. I was opening my eyes back up to just life. And it was just the most reassuring thing, that as weird as it seems, when these really big things happen, nature continues,' she tells me. 'As angry as you can be, that how dare it continue when everything seems so big and impacted on my life, when you realise that they do it's a really positive thing.' Nature, Caroline says, takes her out of herself and puts things in perspective. 'You can't stand alongside a tree and ever think that your worries matter, because that tree has been here for hundreds of years and seen much more things under its boughs than you've ever seen. It just gently reminds you that there's a bigger order, and on it goes, really.'

The kettle boils as she talks about her children, her childrearing. Her girls were brought up among the restoration of the house. 'I wasn't one of those parents who was all "Let's rattle this and sing along with that",' Caroline tells me. 'The house became part of their lives, and then the garden when I was working on that. And that was a really good thing, because from a very young age they got a sense of me other than just being "Mummy".' I am heartened by this, and tell her so. It feels like a permission-granting, a kind of relief that I could be a mother who has creativity beyond the needs of a child. It also makes me realise my own upbringing was similar. I have memories of summer holidays spent waiting for paint to be mixed, of embracing my mum through her decorating clothes as she made the house as she wanted it, covering the walls in her own determination. Caroline dunks tea bags, apologises for her oat milk. 'It's really important,' she says, opening up the compost caddy and chucking the wrung bags in, 'for children to understand that their mother is a person with needs.'

I couldn't tell you why Caroline had gardened, because it would be akin to telling you why she had done anything in her life; it was part of a broader way of existing that made sense at the time. And yet she had offered me a deep insight into what her garden was to her without prompting. I saw it as a backdrop, a stretched canvas to her life. The garden had been there when she moved in with a young family, it had been there when her husband was killed, it had been there as she fell in love. When we met, Caroline intimated that she might be leaving this place soon, but I didn't get the sense that it would be a wrench for her to move on from a house and garden that had cradled so much life; she would go on, make a new space for herself.

Vanessa Bell is known as an artist, as a matriarch, but she was also a space-maker. Roger Fry designed that garden, Duncan Grant and her son had their say on its creation, but it was a space of her making. The garden at Charleston transformed from slurried farmland into flower beds as she raised her children, carved out her practice, grieved her son and her sister and the men in her life. Without her, it grew long and languid.

I wonder how Kayla is getting on since our last meeting. I hope she has found her space, a new one beyond the limits of the prison and its glasshouses. I wish for her the planters filled with flowers, somewhere with boundaries enough to make her feel safe, with enough scope for her to fashion the next stage of her life. I hope she has a garden.

12

ÆRØ

I N 1975, YEARS BEFORE DIANA broke ground on her garden in Clapham, she and her sister went to the Chelsea Flower Show together. Diana wasn't gardening by that point, although her sister had a plot. When Diana admired a show garden, saying she'd quite like a little patch like that, her sister dismissed her naivety. 'That's not gardening,' she replied. 'Gardening is war.'

'And it's true,' Diana told me, nearly 50 years later, as we stood in the garden she finally got hold of. 'It's absolutely full-on war. I do potter about, now. And I'm sort of more at peace with myself. But when I started, of course, I was very aggressive and very ambitious.'

I see more of Diana as the seasons turn. We become friends, exchanging emails at dawn, the quiet time when both of us are sitting and looking at our gardens. Her garden transforms over the year. On the first morning in August I cycle over under ominous skies. The low, dark foliage of late winter has been swallowed by a riot of growth. A small meadow of plume poppies, pale and feathery, and scarlet crocosmia has risen from the bed behind the house. Hardy geraniums tussle with dried allium heads. The surrounding trees have fluffed out with foliage. In the back, the hostas are enormous and flowering in perfect lilac, not a snail bite to be seen. Astrantia jostle beneath light pink hydrangeas, *Alchemilla mollis* creeps over the lawn. Diana's garden was beautiful in the winter in the same way sunlight through glass can be: pure and mathematical. Now it is a place of pure extravagance, rich with the smell of wet earth. 'Most of my gardening is editing these days,' she tells me, gently chastising the feverish monster she's created. In either season, in either state, the garden holds a gentle air: something distilled and measured, bordering on the sacred. As we walk up the steps, the damp, bruised-purple fronds of the Japanese painted ferns lace our ankles with rainwater.

Diana also appears to be occupying another season. No longer tucked up in a hat and coat, she is taller than I'd remembered, her light brown hair curled neatly at the back of her head, as if the summer has also brought her up and out. We have tea, speak about gardening and writing and the climate catastrophe from her conservatory, as more showers roll through her garden and she urges me not to help with the drinks as Bellamy, her cat, tries to get at the butter on the side. She asks me how my conversations are going, what I've found out. I reply that lots of things are coming up, but control seems to bubble beneath the surface. 'Ah, control,' Diana says, 'yes.'

Control had trussed up Diana's garden when she was struggling in an unhappy marriage; gardening this space had taught her how to let go. Sui had decided not to embark on the 'absolute war' of trying to contort the wilderness that encroached on her garden, but revel in what happened when she observed, instead. Megan had found an outlet to process what she had experienced in growing seedlings. Control had wrapped around my own relationship with gardening. In my mid-twenties, the envisaged shapes of my life twisted into something unrecognisable. Looking and tending to plants made sense – more than drinking, more than partying – because when I did I could see something bigger. I had little control over where I would live or what the shape of my life would be, but I could understand how the seasons worked, I could find meaning in the rise of a green shoot from the earth.

Engaging with the outdoor world, tending to it and helping things to grow helped me to be patient and helped me to accept what I couldn't change, but I still regularly wrestled with the confines of womanhood. Before, gardening had been a kind of remedy. Now it was ingrained into my life. I had built it into my days; it occupied my thoughts. Much of what I did out there was quiet and solitary. At a time when I was connecting my life with someone else's, learning to live together, rather than alone, and making life-changing plans for our future, I nevertheless retreated to the garden to make something of it I was still trying to understand.

In shifting from having my own space – at Treehouse, with the balcony – to a shared one, I bumped into compromise and strange, tacit agreements. I found myself stepping into a role I didn't want to inhabit, becoming a kind of Rolodex of family birthdays and weekend commitments and knowing who insured the house. Our relationship is a balanced one, but I still found myself taking on 'women's work'; those silent, stealthy actions of domesticity. Among the hundreds of answers to the survey I'd sent out were a couple that stuck with me for their brutal honesty: women who said that they were drawn to gardening because they hadn't any other choice: 'My husband has no interest in it so it is left up to me. Due to work commitments last year it has become totally neglected and I am determined to conquer it, albeit on my own.' So often gardening falls under women's jurisdiction in the house simply because it won't otherwise get done.

In recent years I have felt my rage more keenly. Anger at the quiet injustices of existing as a woman, of earning less, of having less time to be visible, of being quietly underestimated most of the time. It simmers steadily. Sometimes when I am tired, or fed up of being asked questions, or frustrated at feeling so stretched between all the different things it feels we must be, this rage grows loud and powerful. I lose my temper, seemingly over something small. In reality, it is an iceberg: so much is hidden below the surface.

The outdoors has always been where I take myself where I catch my breath and I let my mind still. The garden has become that too. It is a place built from my control. I alone decree what is planted and where it goes. What emerges from the beds is in a constant dance with what I had imagined. Matt doesn't have a say because he hasn't asked for one, and he is relaxed enough not to notice or mind. The result is a unique space in our home: somewhere fashioned and shaped by just one of us. I suppose it's been a backdrop to a kind of independence, a plot where I can navigate the new changes in my life.

•

As the ferry pulls in, I look through the salted windows to see who's waiting. A couple of elderly women buttoned up against the wind; another in her thirties. Of those standing there, she's the most likely to be Camilla, a photographer I've never met before, but she looks too drawn. I have spent the day on trains and buses from Copenhagen to arrive here, on one of Denmark's smaller islands. The locals dismount, find their cars and find their people, and for a few moments I am the last waiting. Then, I catch my name on the wind and see her: tall, broad, a long arm waving. Camilla, in yellow boots, her rangy daughter at her side.

'You can always spot the newcomers, they tend to look lost,' Camilla explains before hugging me tightly. We are two strangers embarking on a weekend together, and I search Camilla's face for features I recognise from the scant photographs of her online. Her eyes are the colour of the horizon at sea and the ruddy blush of her cheeks rises beneath them; the longer hair I'd expected has been cropped like a boy's. I'd been aware of her work for several years: Camilla documents life on her small family farm – horses, chickens and, most of all, the garden. While her photography is beautiful, steeped in sea mist and the seasons, I admire it more for its storytelling. Through Camilla's photographs, I had learned something of the long, dark winters here; the quietude of the island; how the soft light changes through the year. When she and her husband took the smallholding over from his parents, they had given it a name: *Sigridsminde*. It translates to 'In memory of Sigrid'. I was drawn in by the idea of a plot of land being tended to by a woman, in honour of another woman. Several months earlier, I'd asked if I could visit.

The skies are flat and grey, the wind constant. The previous week had been the final blast of the good summer Denmark had this year. Even in a jumper, boots and waxed jacket I feel fey and underdressed – the locals here wear rubber boots and thick layers, scarves tucked into their collars: my London softness is showing. After the short drive to the family farmhouse, Camilla shows me the guest room, which is beautiful. On the bedside table sits an enormous jug brimming with dahlias – fiery oranges and soft pinks. 'I thought you ought to have flowers,' Camilla says, a little shy, and it all feels so much to be taken in so kindly.

We tour the garden in the last moments before sundown and it's rich with the sweet rot of early autumn. Windfall apples lie underfoot, those still attached canker in the tree. It is a heady reminder of my favourite time of year weeks before it arrives in London. The garden is newer than it looks; the hedges have a slim adolescence about them, clouds of cosmos and dahlias distract from the absence of mature shrubs. This is space that Camilla has created for herself only in the past couple of years. Before, the children played here. They're a little older now and have other claims on the space – a paddock, a couple of football goals. She will tell me that part of her feels she expelled them from the garden a little early. 'As a mother, maybe I would have waited a year more. But I know it was necessary, and they don't really care when it comes down to it. I think I felt so encroached in my personal space, both mentally and physically, for so many years, that I thought, "No, actually, that's okay."'

To me, Camilla's is a gardener's garden. Old steel watering cans punctuate the paths, the wheelbarrows are full of faded *Echinops* globes, topped with a single gardening glove – for grabbing – and a pair of secateurs, for cutting. Flamboyant dahlia heads, the size of dinner plates, are left in a line of rotting plant matter a few feet away from the beds themselves – a clever and colourful slug barrier. There are chairs – often metal, sometimes wood – scattered around in different nooks: apparatus for a garden that is sat in, alone, and observed. Whatever rustling the polytunnel might be making is drowned out by the poplar trees, which hiss in the wind near-constantly. A few minutes' walk will take you to the shingled beach at the edge of the farm. It was here that Camilla and her father would shore their sailboat when she was a girl, and enjoy the quiet spot, with no knowledge that her future husband – her future life – lay on the land beyond.

It's funny, the intimacies you can share with a stranger. Camilla and I hadn't met before, so neither of us knew what we were like with anyone other than ourselves, although she would pause our conversations to speak quietly in Danish to her children and husband. Over the course of two days we dive deep into the stuff of who we are: what it is to create, to put your work onto the internet, to make the choices that shape a life. We both ask frank questions and give honest answers. At some point, perhaps sitting by the still waters of a beach, or perhaps the night before, as dusk quietly closes around the island, we forge a friendship.

Before meeting her, I heard Camilla speak about *Sigridsminde* on a podcast. She spoke about her life: her mother's unstable behaviour that dogged Camilla's childhood and adolescence, her career as a jet-setting wedding photographer, the mental health breakdown that rooted her back in *Sigridsminde*. On the podcast, Camilla spoke about her garden, about her interactions with the outdoor world, as a way of regaining a balance that had been lost. Before I met her, I imagined that this might be the story she told me: of illness, of retreat, of a practice rooted in her garden.

Over the course of the weekend some of this softly bubbles up. We chat about the confusion of late adolescence; a discussion about other places is tethered by her admission that going abroad was trying for her own mental health. She speaks about the darkness that has beaten a steady rhythm in her life with the ease of someone very familiar with it and at times it sits between us – in the car, at the kitchen table – like an old leather holdall. But I never probe. I don't see the necessity; there is far more about Camilla and her life that I am interested in.

She shows me around the island – small, and somehow flatter than I had imagined. We spend a few hours apart and, later, I find her in the garden, tidying up the detritus of late summer growth around the tomatoes. The poplars shake in the wind; it is cool enough to warrant a jacket. We sit down on metal chairs where the hedges make a corner, and I switch on the recorder.

'Sigrid was the woman who lived here when my father-in-law was born,' Camilla begins. 'Her husband was a very ugly human being – on the inside, at least. I guess part of her story is her fight not only to survive on a small farm, as you had to at that point – it wasn't an easy life living in the country back then – but also married to an unkind human.' Camilla tells Sigrid's story with the rhythm of a fairy tale. The plot where we're sitting is situated just a few hundred metres away from where her father-in-law was born, one of many children to 'overwhelmed parents, poor parents'. Sigrid and her husband, meanwhile, had no children. Her father-in-law found a kind of escape here, helping Sigrid and her husband out on the farm before, when he was old enough, moving in and working the land. When the unkind husband died, her father-in-law took over. 'When I listen to him talk of her, and listen to him talk of his biological mother, it's clear that a family can be many things and not always a blood thing. Because even though he is full of respect for his mother, his emotional attachment is to this place, and to Sigrid and not his biological mother,' Camilla explains. There's a photograph of Sigrid, she says, when she was about 18. 'There's a lot of kindness in her eyes.' When Sigrid was old and being cared for, she would walk across the field that separated the farm and the hospital and back at night to feed the hens. 'None of the hospital staff knew,' Camilla laughs.

Camilla holds Sigrid as a kind of totem, in equal parts cautionary and admirable: Sigrid was a survivor, but she also had to be. The fact Camilla's children are so integral to this place is something she considers with Sigrid in mind. 'My father-in-law is not a crying man, but every time he sees the children's lives here, he gets teary-eyed. So it must have been a big issue for her that she wasn't able to have her own children.'

Over the past 12 years, Camilla and her husband have transformed the smallholding – 'I think if she was to be dropped back alive on the planet and put right in the centre here, she wouldn't know where she was,' Camilla says – but it is Sigrid's name carved into a wooden sign on the fencepost. 'Her handprint is here, and on so many things, both outside and inside the property.'

But Camilla's handprints are also all over this space. What I hadn't realised before I met her was that she moved to the island when she was barely out of her teens; not much older, in fact, than Sigrid was. 'I knew that this place was part of the deal. If [my husband] was going to be my partner in life, and this was also going to be our place in life,' she explains. 'The plan was to move to Brussels, work as a translator and live in a fancy apartment,' she laughs. Instead, she fell in love with a local boy, and settled in the clutch of houses where he and his family had grown up. 'You can only live one life,' Camilla explains, with the frankness she deploys for many things. 'You don't really know where other paths would have led you.'

The transformation of the house has been a physical, financial and intellectual challenge over the past 12 years, but also an emotional one. 'It's been hard-won,' Camilla says, adding that a decision to move the front door was a six-year-long battle. Even without much explanation I can understand it: Camilla's home is beautiful, in a very considered and comfortable way. She was a girl coming onto the island, wanting to make a home for herself in a place laden with the histories of her husband's family. 'I really felt out of place for a while and then I came with all my ideas and all my creativity and all my initiative and ambition, and yeah, it was just a clash of two worlds, kind of,' she sums up.

Rather than the island being a retreat from a glamorous, jet-set career – as I had originally assumed – it has been an isolating force for Camilla. As we drive around its gentle slopes and sea views, she speaks of Jutland, where the landscape is more dramatic. We pass a chic modern timber-clad house – a world away from the heavy walls and cosy low ceilings of *Sigridsminde*, which was built in the 19th century – and she points and says that's what she wants to live in, one day. 'I had a phase of about four or five years where I really resented being here and I really felt trapped,' she tells me later. 'I felt like I made a mistake: we never had any debate, like, "Where would you like to live in the country?" It was this. And yeah, I think the negotiation part of it sort of wore me down.' Over the years, she says, 'I sort of found my way of creating here. It does feel like home now.'

While the garden is integral to the slow, seasonal pace of life Camilla and her family live by – eggs from the chickens, vegetables from the garden, meat from animals hunted by her husband (something far more common in Denmark) – it's less of a retreat than a muse for her. I ask if she's able to go outside when she's mentally unwell and she shakes her head. 'I really wish I would. But no, I do have to sort of scrape myself out of the bottom and be on an upward trajectory before I can be here,' she replies. 'I never like to work out aggressions or anger or sadness here. So no, when I'm feeling really, really bad, I usually stay indoors. And then when I start to be able to breathe a little bit again, then I will go out and then it will help me carry on.' This challenges my assumptions. Camilla writes about her mental health, but I'd chosen to believe her garden, this island, to be a restorative retreat. I'd been so hasty in constructing the narrative I'd wanted to see from Camilla's work: the dreamy sunrises, the plot–to–table dinners. In doing so, I'd glossed over the reality of women's work, ignored the difficulties that come with the choices we make as women.

She gardens, Camilla tells me, in the same way she once 'created with paper and pens'. Home was unstable when she was small; the family moved around a lot. But one constant was her grandfather's garden, a place she still describes excitedly. 'It was full of tunnels and the trees all reached each other and there were small paths. It was really a garden for discovery. I think everything I'm trying to do here is not so much the plants but to see if I can recreate the feelings I've experienced in my grandfather's garden. It's not so much about the physical thing, but more about how they make you feel.' Suddenly, I can see the expectation put upon the trees and hedges that surround us, I can catch a glimpse of the image in Camilla's head of what she wants the garden to be. 'There's still many years to come here, before that vision is sort of complete,' she says. 'And it requires constant work. It's not like a photo or a painting where you do this, you do this, now it's done, and you do whatever you need with it. This is a constant work in progress. It's a completely different way of creating – both exhausting and exciting in a different way.' I ask if it will ever be done, if she'll know when she's reached a point of satisfaction. 'At a certain point, the hedges will have grown to the height that I envisioned them, and that will be a point to pause and reflect and look back at what work has been put into it,' Camilla replies. 'That will be some kind of finish line.'

Sigridsminde is Camilla's home, but it's also a legacy –

and a heavy one. She explains that her children, not yet teenagers, are insistent they want to live here forever, just as her husband took over from his parents. It's a notion she presents with a kind of sigh. Camilla and her husband have created a childhood far more blissful than hers ever was, but it also means she must stay here. Just because something looks wonderful doesn't mean it can't be constraining. As we drive around the island, she is honest about how it bores her. Her next garden would ask even less of her – a vegetable plot, surrounded by the wilderness. With every passing year, she releases a little more control over the garden at *Sigridsminde*. 'I think it's just a culmination of everything,' she replies, when I ask if the garden has helped her to accept herself. 'Gardening, motherhood, being in a relationship with a man since you were a teenager, growing up, having to still figure out how to navigate with another person so close to you,' she lists. 'The garden has been through several stages before it has become as wild and unkept as it is now. Each step has been less perfect and less immaculate. So in that sense, I have also become less concerned about being perfect myself.'

I wanted to come to Ærø because I wanted to see a garden that was indebted, in some way, to another woman. What I found was a plot of land that held – in a beautiful, near-invisible silence – undulating stories. Sigrid and Camilla's lives are not the same: Camilla has a good and sturdy marriage, together she and her husband have created a home for their children that is unconstrained and immersed in nature; their childhoods have been stable and free and safe in a way hers never was. But both came to this place as young women and carved out their own lives against the grain of expectation. I realise how naive I was to think of *Sigridsminde* as something as straightforward as a halcyon, back-to-the-land escape. 'This place,' Camilla says, 'has been just as much a source of stress and conflict for me, in my marriage and in myself, as it has been the salvation.'

I had been looking for straight lines when the truth is far more gnarly. With every conversation I'd sought a causation, a story with a beginning, middle and end: that something happened, and then she gardened, and then she felt a certain way. I think I craved this simplicity because I thought it would offer me clarity. A means of revealing that I gardened because it was in my blood, or because I was escaping the confines of what was expected of me, or because it gave me a sense of control when I felt I was losing it. I knew that was too simplistic but I still expected it of others. Gardens, as with lives, are not so straightforward. I had chosen to see Camilla's as idyllic, then I saw it beyond the flat surface of a phone screen and understood that it was the product of work, of years of fighting for agency and control. Women's lives are arguably the same: we are raised to make it all seem effortless, every small choice that we make. We must not boast about our ambitions, our successes, the beauty we create and uphold and fetishise. We have been taught that to lose control is to lose. So when we choose to let go in the garden – to let the weeds flower, the grass grow long, the pests multiply – it is a far bolder choice than it may seem. We are showing that we are doing as we please, that we are ushering in the wild and seeing what happens. We are reclaiming beauty as something tumbling and unexpected, rather than the tight, shining standards we have been raised with. We are making the wilderness our own.

•

The weather app tells me the sun will set in 10 minutes. I pull on a jumper, thick socks over my tights and my overalls over the top of that. There are bulbs to plant and I am restless. It is mid-autumn and heavy with it. The clocks changed a couple weeks ago and the dark after-noons have been clawing at me, or perhaps I have been clawing at them. Even with a diary jostling with invitations, I feel detached – from people, from the seasons, from what our lives look like now.

The garden shows markers of change that are more difficult to trace elsewhere. This time last year, the ground was bare, the beds freshly cut, one corner of the lawn was a boggy mess. The garden I have made has turned out to be a late summer one, full with mallows, dahlias, persicaria and fluffy grass when others are waning, and I expect it to fade more quickly. But even with tufty winds and bucketing rain, it holds. Fat female spiders leave webs for the dew to cling to, and I am proud that they have come here, little symbols of worthwhile ecology and maternal sustenance. One Friday morning, I take gloves and secateurs and muscle and clear the earth. Nasturtium tendrils; scraggly cornflowers never to bloom; fading, mollusc-munched dahlias – they all go in the barrow. I hack back the artemisia, uproot some suckers for cuttings. I rake leaves with my hands. I pile it all on top of the compost bin, far above my head. I leave towering seed heads and tell-tale stumps. The bare earth I am left with looks like grief and opportunity at the same time. It is here that I dig, nudging my spade between the perennials, to make troughs. In my grandmother's basket, daff and tulip bulbs lie waiting. I push my hands through them, mixing them up, then grab rough handfuls and press their tufted bases against the soil.

In the garden things shift and float: boredom, loneliness. When I sow sweet peas, when I dust off the coldframe, when I plant bulbs, I think of the woman I was a year ago. I look at the garden that she and I have grown. It isn't that I'm necessarily happier out here – although I often am – but that I am able to better understand my purpose. There will always be to-do lists, but increasingly I ignore them. Last autumn, I stared at the bare earth with impatience, willing it to grow. Now I peer and stretch into the beds, unearthing the scalloped edges of poppy leaves, feathery fennel and straight sweet pea seedlings. Beneath the grasping ivy lies an entire flower bed that has waited for its emergence. One rainy Saturday morning I uproot ferns from a waterlogged trough and find them a drier place to settle. The rain pelts, but I've stopped noticing it. Out here, beneath open skies and vapour trails I can fit myself into a whole. I used to talk at the garden, now I listen. Sometimes in the house I am still too busy trying to fit myself into a mould of womanhood I don't recognise – the accomplished cook, the proud homemaker – but in the garden I can be and do as my body tells me, as my mind suggests, wrist-deep in the soil.

13

CAMBRIDGE

O NE EVENING I MEET WITH an artist named Elaine, who has a wide, open face and auburn hair, and swiftly sets upon pulling a kind of picnic from her backpack: bags of crisps and cans of beer and a bottle of wine – just to give an option. She tells me she will turn '56 or 57 at the weekend, I can't remember' and, in the tender manner with which she speaks, that she has 'had quite a bit of life'. As we talk she scans through it, casting back to being raised by a woman who was always 'getting the camera out at those traditional moments' but nevertheless did it well and sparked a fascination with photography that would become Elaine's career. She started as a photojournalist in Afghanistan, Palestine and across Africa and has ended up, now, making 'deeply feminist' work with plant matter at the heart of it.

In Elaine's eyes, all manner of things became specimen-worthy. Vintage stockings and twist-ties, pebbles found by the Thames at low tide. She says she 'had to learn what an artist is, and I had to learn how to call myself an artist. To own that title.' When Elaine moved in with her current partner of 10 years, after both had left long-term relationships, she did so into a house with a garden: something she'd never had access to before. In the garden was an *Acanthus mollis*, or bear's breeches – an imposing and handsome plant that will happily take over the corner of a shady garden. Elaine didn't know what the plant was but was fascinated by it. She would watch what it was doing on a daily basis and cut away parts of the plant to scan them – a key part of her process. 'At some point in the summer, we would sit outside and I'd always say I could hear things, like, exploding,' she says. 'It took us weeks to realise what it was, these loud popping explosions. And then I finally realised that the seed cases were literally cracking open, and not only that, but they were flinging stuff out quite some distance. I looked at the language that was used, and these freaking things are called ejaculators! I realised that there was all of sex, all of life, in this thing,' she says. 'It was a sort of weird revelation.'

It collided with much of Elaine's other thinking around her work – work that has been about looking closely, and thinking about language, and what we miss in plain sight. 'As women,' she continues, 'we have always been associated with plants and flowers, and we've been held back because of religion that defines us in terms of the innocent and the erotic. You know, how we get *deflowered* – how powerful is that? I just haven't stopped thinking about it or making work to do with how plants and flowers have kind of shaped how we, as women, have been seen and perceived in so many ways and how it continues.' So much of this is ingrained, I realise. Mary Delany built sapphic desire into her gardens in the 18th century, but we've been sanitising our flower beds while cooing over the floral sex organs on display ever since. Gardening is grubby work – we sweat, we bruise, we get filth under our finger-nails – but women are expected to do it politely, sweetly cutting flowers while erasing the effort that went into growing them.

There's a near-silent history of power and an ownership, Elaine goes on, to the images women artists make of plants. We talk about Georgia O'Keeffe, and Imogen Cunningham, an overlooked pioneer of 20th-century photography who shocked onlookers by photographing her nude husband in scenes of nature as early as 1915. Cunningham was diverse in her subjects – she took self-portraits and close-cropped nudes and street photography – but, not unlike Elaine, she began to take photographs of plants and flowers because, in her words, 'I couldn't get out of my own backyard when my children were small.' The photographs from Cunningham's garden are wrenched of any sentimentality about early mother-hood: the contrast is stark, the plants are treated almost anatomically. Cunningham's dramatic and abrasive images of calla lilies and magnolia flowers abstracted the conventional understanding of botanical photography. Male contemporaries, such as Edward Weston, and even those who came later, such as Robert Mapplethorpe, are much better known. But even at the time Cunningham resisted being defined by her gender, insisting that she was 'a photographer, not a woman'.

'But there was something in those images that very much defines the new woman who was much more confident, who was having shows of work and was a bit more likely to be seen for her own,' Elaine says. 'Those pictures are really much more important than people realise, I think, because they symbolise a kind of independence and confidence in their work.' Cunningham didn't have it easy – she was frustrated with her gender eclipsing her talent at the time – but she created her best-known work while trapped in a realm of domesticity as the mother to three little boys.

As with every other woman I've asked to speak to, Elaine chose the location for our conversation: Red Cross Garden, a small, almost cottagey park within reach of London Bridge, which has a tiny bandstand and a pond. We're there in the last hours before the keeper will lock the gates, and everything is still lazily in flower. On one of the redbrick alms houses that surround it is a blue plaque dedicated to Octavia Hill, a Victorian social activist who designed it. 'She's a fascinating woman; she was a feminist without realising it,' says Elaine. Hill was the only woman in the trio who founded the National Trust; she took poorer people on nature study walks around London's common land and rallied for better social housing. She understood that 'we all want quiet. We all want beauty . . . we all need space. Unless we have it, we cannot reach that sense of quiet in which whispers of better things come to us gently.'

I'm not sure Elaine and I are doing much for the quiet of Red Cross Garden: we make one another laugh, we enthuse over unexpected points of connection; while she is softly spoken, I am not, and I marvel at the way she explains her work and her life. After a while, we switch off the dictaphone: Elaine wants to chat more informally, and I let her. With a new firmness, she imparts a kind of maternal wisdom from what has been an unusual life – earlier, on the record, she said that she was brought up in what she calls 'the cult of the Catholic Church' but has since rejected all religion. Elaine became pregnant at 30, at the same time she was planning a 'big career move'. It wasn't something, she tells me, she found easy, but 'it turned out to be the best thing ever'. And so we talk about creating things and raising children and I admire and appreciate her honesty. A chunk of my creative life, I learn, will go on hold if I have children, and if that happens, it will be fine.

There was a collision in all of this – the looking closely, the warmth, the kindness and insight from a stranger, the secret confidence women communicate through their work – perhaps not unlike the cracking pop of an acanthus seed pod in late summer. I was gaining strength from a growing certainty and from being brave enough to be honest with myself in realising what I might want from life. Life could change in a heartbeat, but it also moulded almost imperceptibly over time. 'I'm not going to try and be clever about it, because I can't,' Elaine had said. 'But I think my experience of the garden was exactly about time, and it was about change, daily change: the miracle, or the mystery of things.' This is the stuff that life is made of, too.

I'd been curious about the community that women formed around gardening ever since I started asking these questions. I searched for groups of women that gardened together and read books about women's land networks in North America, where groups of often queer women would gather and take over land to exist in a society free of men; made enquiries to visit farms run only by women. Women have gardened together for centuries, just as women have gardened alone, but both have often gone undocumented, buried in the invisibility that wraps itself around women's domestic work. I wanted to know how working the land together affected female relationships. What I hadn't expected was to forge some new relationships of my own.

Elaine sparked connections – between Octavia Hill, Imogen Cunningham, the silent audacity of a flower bed and her own creative process – and speaking to her made me realise I'd been doing the same. With every conversation I'd deepened my understanding of my own relationship with the earth, and of the different ways women grew.

These women were alluring to me because they had lived lives that were different from my own. I caught glimmers of all the lives I hadn't lived but might yet still – of divorce, and second loves and starting again, of moving abroad and staying put. Gardens tethered people, and they helped them ground themselves in new places. I suppose I was trying to imagine what might happen if the marriage I was embarking upon didn't work out, or if I could give up the stability of the life I knew – nine-to-five, mortgage, a city I'd inhabited for a decade – for something less familiar. These women didn't provide a how-to guide so much as a reassurance: that the ground would hold us when we needed it to.

•

In the garden that Ali and her partner Sarah tend to, a rose named 'Compassion' grows from concrete. It has soft, coral-coloured petals, and was planted by another Sarah: the woman who owned the end-terrace house they live in and who gardened – ferociously – the snug plot that winnows around it. Ali and Sarah (Smith and Wood, the novelist and filmmaker, respectively) had been neighbours of Sarah's before they moved in: in an act of safekeeping, she'd given them first refusal on her home after she died. In the late gardener's final years, the women would carry out small tasks for her, such as sowing seeds in a trench by the side of the house, and Sarah would deliver them with flowers. 'She'd come to the door with roses and a little note – I've kept some of them – that would say: "Ali, it's called 'Compassion'". And then, at the bottom, in her beautiful writing, "Thank you for watering".'

'She laid down plants for life against the fact that she was very ill,' Sarah explains, 'and with great symbolism.' This gardening benefactor crops up often during the conversation I have with Ali and Sarah, on a warm late June day, in their garden in Cambridge. She was a terrific gardener, they say; her decisions, her planting, her creativity still unfold in her absence. Tucked behind the house is a shade garden, near-impenetrable due to the prolific flowering of 'Bobbie' – short for 'Bobbie James', a white rambling rose that acts as a kind of seasonal security guard. Beyond lies Sarah's legacy: a deep, lush corner of fern fronds and foliage, growing in a green-slicked cave edged by jasmine and honeysuckle. Small and shady, it's an easily overlooked masterclass in low-light gardening, as well as a good place to sit. 'It was a gift after she died,' Sarah explains. 'We missed her but then every month something else would arrive. It never stops flowering in some dimension, even in winter it's got beautiful things coming out; it's like a calendar of gardening. It's a real gardener's garden.'

I'd asked Ali if we could talk about her garden after I read about her compost heap in *The New York Times*. She was among 75 creatives asked about how they spent their lockdown; her and Sarah's compost heap was the response to the newspaper's question: 'What's one thing you made this year?' (When asked 'Did you have any particularly bad ideas?', Ali replied: 'Putting the compost heap so close to the kitchen windows'.) I'd also been taken in by *Seasonal*, her quartet of four novels written almost in real time between 2016 and 2020; to me, they were as much reflections on the natural shifts in the year as the increasingly unfathomable events of the news during the last half of the 2010s. Ali agreed to speak to me, but denied that she was much of a gardener, and suggested that our conversation included Sarah, who was.

I find Ali before I find her home: she sees me turning down the hidden narrow street on which she and Sarah have lived for 20 years and waves an arm, the other holding on to a pair of loppers. She's wearing wide-legged, patterned blue trousers that lend her a stature that belies her height. We walk around the garden and it becomes immediately apparent that Ali is more of a gardener than she lets on: the roses are pointed out, so are the apples. These are her two great horticultural loves. The loppers are closed and placed on the ground. We come to a stop outside a low, one-roomed building – a sun-filled hut that emerged during lockdown – where Sarah finds us. Like Derek Jarman's Prospect Cottage, the hut is black with yellow window frames. It was not easy, Sarah says, to match the colour.

We sit in the hut. The large doors are open and through the windows you can see the trees. It feels very much as if we are both inside and separate to the garden, as if we are occupying another space within it. Sarah, who is considered and a little wary, with large, brown eyes, comes from a thick rope of women gardeners. She describes gardening as something she assumed she'd do rather than chose to, a means of repeating the actions she learned through watching – her grandmother swiping plants from stately home gardens with the spoon and carrier bag she kept in her handbag; of being a small child, standing behind a gardener hoeing. 'It's all sort of memory,' she explains. 'You kind of know how to do it because you watch someone else.' It's slippery as a result. Sarah's mother was involved in the renovation of the gardens at Whitehall, a historic house in Cheam. Recently, a historian has been in touch to find out about it, as the work of those who rescued the place was never recorded. 'She planted a herb garden, and she learned a lot about what to plant from my grandmother,' Sarah says. 'It was a new conversation for them to have, her and her mother-in-law. It was a funny thing, like a secret conversation. My grandmother had inherited the folklore, it was all there. You can research things in books but there were people really doing it just out of habit and assumption, and passing on stories to each other. It's almost a secret language.'

Ali traces her love of apples to her father. 'When my mum died and my dad moved out of town, he moved to a small house in a tiny town on the Black Isle, just north of Inverness, and he filled his garden with apple trees, about the same time as I was planting mine,' she says. Ali's father espaliered his trees, lining them up in rows. 'He loved his apple trees and I didn't know he loved apple trees, but I love apple trees and I love apples,' Ali explains. 'I love the simplicity of it, and the natural generosity that they are.' Her mother would plant rowan trees outside the front door – 'She would not live in a house that didn't have one at the front door' – and Ali remembers 'the most leafy window, the swaying of fresh green all the time against the bedroom window from the rowan tree that they planted when I was very small and grew to the height of the house'.

We are talking about inheritance, but Ali and Sarah have made gardens together for years – ever since they met as students ('See that was *one* good thing about Cambridge,' Sarah quips). There was the house where they lived as graduates, where the landlord was kind and would accept whatever rent they could afford, with a fuchsia and a peony bush by the door. There was the place in Edinburgh, which wasn't so much a garden as a depository for the plants they had taken up in a van that was stopped by the police for being overloaded and snuck under the weight limit only to break an axle in York. 'Two weeks later, when we came back to collect the van, we found the mechanic had watered all of our plants,' says Ali. The garden we are sitting in was next; it too has a storied history, swelling from a roof garden to the generous, tree-laden plot I have been shown around. Over the past two decades the women have taken a cautious custodianship of it, renting the land from a neighbour who is in his eighties. 'There's the constant sense that it doesn't belong to us, this garden,' Ali explains. 'And yet at the same time that it is contingent makes it even more wonderful that we have it. Imagine planting your apple trees and knowing that someday someone will say, "You can't keep your gardens." You're aware of it all the time, how precious the space is.'

They've come to cherish the garden more over the years, Sarah tells me. 'I don't know if I would always have thought that having quite a lot of outdoor space was imperative,' she says. 'But now I'd rather have a garden than a house.' What, I ask, would they do if it was taken away? 'We'd form a human chain around it!' says Sarah, not entirely in jest. 'Our plan, if they decide they're going to sell the space, is to mortgage everything we've got and buy it,' says Ali, at the same time. 'And if that fails, find a garden. That's what we would do. Because we couldn't live without this. We could not. It's such a wonderful thing to have.'

Ali introduces the roses to me by their names – so often with roses, I remark, deliciously kitsch. Over the years, she and a neighbour who shares the garden have bought them for one another: '"Grace Kelly" or "Judy Garland", "Elizabeth Taylor",' she hoots, her Scottish accent stretching the vowels. 'There's an "Ancient Mariner" over there as well. But that one, that one next to the ash tree, is a "Shakespeare".' The 'Shakespeare' rose is the most special: a deep, full cerise old rose with a potent scent. It has also been discontinued by the nursery. 'You can't get them any more,' Ali says, with a reverent hush. The one we're looking at is 20 years old. 'It was a bare-root that just lay at the back door for months and then we put it in, and it just bloomed and grew and bloomed that first year,' Ali says, still amazed by it. 'It's never stopped giving the most sumptuous, beautiful, casual roses.' I wouldn't have had 'Shakespeare' down as a flower so pink and frothy, I tell her, and she laughs. 'Would you believe it! But you know, it's like Shakespeare in that it's holding out against expectations. I mean, it's given and given – you can't get them any more and there it is.'

Later, we turn to words: we try, collectively, to think of women in books who were gardeners and we struggle.

Ali explains there's another rose in the garden, named 'Young Lycidas' after the poem by Milton, in which the poet mourns his friend by denying their death. 'I associate the brevity of our lives and the determination to last beyond the gravity of our lives, one way or another, with roses,' she says. They are a plant, she explains, that exist long beyond their brief flourish. 'Whenever I think about roses, I have in my head that there are next year's roses, and the year after that. Roses always reach ahead.' Gardens do, too. I think about Diana, her comments on how old gardeners stave off mortality by thinking about the garden; I think about what we leave behind in a space. This was a garden created by another woman, and held close. When we make space, we do so knowing we will inevitably leave it.

Sarah doesn't like roses, she explains. She's had to 'give in' to Ali's affections for them. 'I don't like gardening with them, it's a nightmare of being stabbed by the thorns every time you have to deadhead them.' They are, like the apple trees – which are wrapped in corrugated cardboard and treated for moths and watched closely for blossom – part of Ali's domain. 'She's a real gardener, a real gardener,' Ali says of Sarah. 'What, because I know the names of plants?' Sarah replies, sceptically. 'No, because you really make things work and you make them beautiful,' Ali says. 'You make something of whatever it is you touch, just come alive and fruitful and fertile. You clear a space and suddenly everything just fills the space in this beautiful way as if you've magicked it out of nowhere.'

It's a lovely thing to witness, this appreciation. Of the dozens of women I have spoken to by the time I meet Ali and Sarah, all of them have gardened alone. I understand this: I also garden alone. But this search – for women, for conversations, for understanding – emerged from an isolation. In undertaking it, I have grown a community.

A shift happens. The conversations begin to still, or maybe they just change. I retrace my footsteps. In the months that have passed my relationships with some of the women I've met have evolved and grown in their own way. We speak online and over text message, we bump into one another at mutual friends' parties. I cut flowers from the garden and walk to Pauline's, where she shows me albums filled with photographs of the houses she used to live in – many of them layered with the textiles she has made – and the warmer, sunnier places she and her friend went to paint. A spontaneous look at a life: the haircuts, the fancy dress parties, the sofas and piles of books. When Marchelle gets good news I call her up and we talk, excitedly, as I walk around the garden. One short December day I pedal over to Diana's house and we sit by the fire, talking about big questions and the etchings on her walls as tea cools between my palms. I meet the baby that has been born since I spoke with her mother. As New Year's Eve approaches, I reflect on what I've done over the past twelve months, and realise that I'm proudest of my compost and the new friendships I've made. Maya and I watch our acquaintance toughen into something hard and good; on the first day of the new year, we text one another photographs of the views from the hills we have climbed. This strengthening network of women I'd never have met otherwise, who have shown me their gardens and told me about their lives, is something I'd not expected when I opened that green notebook and started writing. I expected to learn about gardens. What I hadn't known was how attached I'd become to those I met.

•

I spend a couple of days with Hazel, this time with her sister Susan, in the depths of mid-Wales. We're on a farm on the remits of New Radnor, a Welsh village so remote that even the local cab driver referred to it as 'the wilds' on the drive over. Behind us stretch fields and hedgerows, and, beyond, the heavy grey thud of distant hilltops. In front reaches Bache Hill, a towering lump emerging from the Radnor Forest. The few solid, white-painted and slate-roofed farmhouses are the only signs of human habitation around. Otherwise, it's just sheep and sky and bluster. Susan and her husband run a low-intervention vineyard on these hills; she and Hazel have plans to collaborate on growing flowers here. They agree to meet me because I was curious about gardening sisters, and I follow them as they paint big pictures from empty plots. Snippets of their upbringing ricochets between them; Hazel was dragged around garden centres with their mother – who left Grenada for England to marry their father and still 'loves the concept of the quintessential English garden' – while Susan was taken to National Trust properties after older siblings had left home. I watch them trace the connective tissue of what they were shown to what they have become. Both women agree that their mother's love of gardening wasn't 'imparted to us directly – it was her domain', but both have found grounding in the earth. 'It's funny,' Hazel says, turning to her sister, 'we've never really spoken about this, have we?'

Once we leave the fields, we end up getting a lift from Susan's father-in-law up the hill that looms overhead. The narrow road twists as the 4x4 rears up it; fluffy conifers are within a hand's reach of the car's windows. It's a spontaneous adventure, one that brings a kind of excitement with it. Susan and her husband used to come up here when they were younger; they've named their vineyard after a rock formation on the side of the hill. I stand back and take photos as the sisters admire the view. You can see for miles from up here, far enough to see the cloud split into two: dark above light, and beneath that an endless rumble of blue hills into the horizon. Among

the cobweb of hedgerows below, you can count the buildings on two hands. Before me, Hazel and Susan wrap their
arms around one another, feeling the push of the air
against their joined bodies. Their hair, their jackets and
scarves are borne aloft. I think about my recorder, about
the shape this interview has taken. This whole thing started
with Hazel – a curiosity pushed into the earth to see
what might grow – and now I feel it reaching a kind of
ending. I have cosseted these conversations as I might a
growing plant: I have looked for women, I have summoned
up the courage to request their time and travelled hundreds
of miles to meet them, I have sat and listened to stories
from their lives to better understand my own and I have
kept their words in mind. The past couple of years I have
felt cast adrift in my questioning: what is it to be a woman,
a wife, a mother? How can I manage what I want with
what I should be? The stages of my life have bumped and
separated like tectonic plates, and it has taken time for
me to find a way across. I have found some passage in
the garden, but I have gained courage from sitting in those
loved by other women. On top of a hill in rural Wales I
look out and see two sisters laughing in the wind and
realise that I have grown, that I have been nourished by
this search for generosity and wisdom. It has given me
the confidence and permission to continue on my own.

•

It's just gone lunchtime in Edinburgh, but there's barely any light left in the day. I'm at the Botanics, as those who know the Royal Botanic Garden Edinburgh call it, meeting Cynthia, a molecular biologist undertaking her PhD here. The Victorian Temperate Palm House, the majestic home to tall windows and towering tropical trees, stands empty in the midst of a seven-year-long facelift. We are caught between things: a moment of stasis and transformation.

Cynthia is an academic and a scientist, but she's also a florist. Her work makes grown things look alien: leaves of black kale become as dramatic as a rippling veil of chiffon; a curly cucumber becomes baroque; a slipper orchid emerges from a bracket fungus like a medieval dragon. Sometimes the plants are in carefully crafted ceramic vessels, sometimes they are in metal cans, sometimes they rise ethereally from cracks in the landscape. Caught between the seemingly separate worlds of science and art, botany and floristry, Cynthia is posing a challenge to these boundaries and making something for herself in the process. We are so often told to pick just one thing to do, but here she is showing what happens when you don't.

Botanists, she tells me, have told her that to cut flowers seems a shame: once you cut a plant by the stem, you've killed it. She doesn't actually come into the steamy labyrinth of working glasshouses to undertake her research at all – 'when I do lab work, the volumes we work with are minuscule – it feels something like mixing one air droplet with another. It's as detached from being a plant as possible' – but she comes in here most days anyway: 'I need to actually be around plants.'

Cynthia's dressed in black, her dark hair a long swathe over her shoulder. We pass through the glasshouses, the air in each a subtly different temperature and humidity, cooing at spotted leaves and thick wads of moss caught in the corners of stainless steel tables. As we walk through the gardens Cynthia talks about her work, more the floristry side of things than the science, quite possibly because I wouldn't understand the latter as well. The two can enable one another: 'There have been a few times when I've been allowed to cut stuff in the gardens and use it, which is literally a dream come true,' she explains. But they also sit in opposition: the surface fascination of aesthetics with the near-invisible study of what lies beneath. She doesn't know which she'd like to occupy once she has completed her doctorate. 'I go back and forth in my mind quite a lot deciding whether [floristry] is what I want to do, or if when I do it it's going to be enough.' Academia, she says, 'is not creative enough, it's not tactile enough. But when I'm working as a florist I get really annoyed that it's just doing gigantic weddings with stupid amounts of flowers that get thrown away the next day.'

We talk about her *ikebana* – the Japanese art of flower arranging – classes in South Africa, where she grew up and began her floristry career. 'When I go to classes with my teachers, they'll say, "We're making this arrangement today. And these are the dimensions, the proportions, the elements." Sometimes, I don't really want to use that thing. I often want to do something slightly different.'

The one thing she enjoys the most, across all of it, she tells me, is how plants grow. 'As humans, when you're exposed to danger you have the ability to run or to hide or to do something,' Cynthia explains. 'Plants can't move. And so they have to be stuck in that position, and adapt to things that happen to them in that position. It's the result of that happening, mixed with their genetics in the first place, that results in them looking the way they look. So like, this plant has grown in this weird way because wind comes from this direction, and the leaves are really red on one side because the sun shines on it. But then also, genetically, it's programmed to express some kind of other phenotype. I enjoy that both from a technical perspective, but also from an aesthetic perspective. I really like what a wonky branch looks like, or like, a mossy stump. I think I was drawn to this PhD because I'm looking at what tells these leaves to grow in a certain way. And I think it's the same with *ikebana*: I really like that *ikebana* likes plants for how they grow.' Cynthia pauses. 'Because for me, it's just really enjoying how something grows and not *needing* it to be any other way.'

One of the central concepts of *ikebana* is *ma,* the Japanese notion of emptiness, or a pause in time, crucial for growth. Without time or space, we can't grow. In *ikebana*, the space around the flowers is as crucial to the arrangement as they themselves are. Cynthia left floristry for academia and has adorned her studies with a creative practice making floral arrangements. To me, plants – her way of observing them, understanding them, arranging and appreciating them – have also granted Cynthia space to find something that accommodates her curiosities. 'It's a very different mindset to me,' she says. 'Because when I cut something and I put it in a vase, I'm appreciating the hell out of it. It's not like I don't care about it, it's like I care about it even more.'

Cynthia makes intricate arrangements outside and brings the stuff of outside – moss, twigs, flowers – in. When she blurs those boundaries, the plant matter she works with takes on an altered, heightened state. We meet in the winter, a time when I have been distancing myself from the garden. In the shorter days it's difficult to find the daylight to get out there, and fewer things need my attention. The previous winter, my time in the garden was ruled by a list of jobs: fences to paint, beds to carve and mulch, wires to be hoisted and twisted and tensioned. But my gaze has shifted, from one forged from determination to something more curious. I mostly take it in from the window, this garden that just months ago did not exist. There are weeks when I barely go outside. Some days it is too dark, too bleak, to go out. I leave much of the herbaceous growth for the insects and the shadows it casts against the wall. The rampant changes that fascinated me six months earlier are slower now. On the internet, I come across photographs of tidy winter gardens, their beds tucked in with dark mulch, their lawns tipped with frost, but mine does not look like this. Overwhelmed by the towering compost bin I open up the bottom and wrestle with what has been rotting in there; within minutes an avalanche of crumbly, good matter lands on my lap. As I shovel it into the barrow and then onto the beds, I am steeped in pride of the things I have made.

Sometimes I worry that my waning winter interest in the garden is a sign that something is broken. I wonder where the compulsion that took me out there so often has gone. I find my curiosity shifting far deeper inside, to the mysteries of my own body. After 15 years of swallowing a tiny white pill every morning, I abandon them, suddenly struck with grief and desperate for a cycle I've not experienced since I was a girl. I am doubtful that anything will change much; I can't believe that a body so rammed with chemicals could learn to function again quickly. But a handful of days into the new year I am surprised by my own blood and it feels like witchcraft. I spend a cold, damp hour outside. I clean and sharpen my secateurs and cut dead things back to the ground. I marvel at the fierce, red nub of a peony pushing through the earth.

In the summer, in Ali and Sarah's garden, I asked Ali about the seasons. The garden helped her to make sense of them, she told me, because it had shown her the seasons in action. She'd seen winter 'clear the rot of autumn off itself as if it shook it off like a dog in the water and then you get that cleanliness, that cleansing of winter'. I'd thought of winter as many things: as dark, as quiet, as a deep and necessary sleep, but never cleansing. But in the stark light of a new year it makes sense. Indoors, I satisfy an urge to prepare – somehow, for something I can't explain – by clearing out cupboards. I see the gradual decay of the garden as akin to this, as a process I cannot hasten or even properly witness. We are both making space for what Ali calls 'the thud of spring'.

The tight, wary consciousness about the next stage of my life that accompanied me into this space has loosened with the seasons. I no longer stand in the garden and worry that I am not capable of what it might demand of me. When I look at it, closely and from afar, it is with more anticipation than expectation. I find myself questioning what will grow rather than when or how. I know that I can't predict its beauty; I know that I will see something familiar and grounding regardless. We have built a home here, Matt and I. We have talked about the future and we have lived in the moment. We have danced in the kitchen and bickered about washing up. We have slept many hours and woken in the night. We have brushed our teeth side by side and shared the shopping bags. We have fed our loved ones and laughed into the small hours. I have stood on the front doorstep in socked feet and stared at the moon.

It is in the garden that Cynthia's words – of 'appreciating the hell out of it' – come back to me, months after she's said them. A faded amaryllis flower has been drifting around near the back door – I've been meaning to deal with its bulb properly, feed it and store it to usher it out of the darkness again later this year, but it has just sat outside, choking in rainwater. Its petals, once full and pink, have become silvery, papery husks. They grace the wet grass of a left-alone lawn. I see them, then compost them, leaving the bulb behind.

Sometimes I get so caught up in the notion of what life might be that I forget the meaty matter of it around us, the conversations up and down the stairs, the drifting light of the sun between clouds refracted through a windowpane. And in these times the garden roots me: it is very difficult to think about what will come when faced with a flower that will only be there for a few days. I garden for many reasons: for solace, for joy, for release, for connection. I garden to force myself to be in the moment, to encounter daily wonder, to solve problems and make good. I garden to remind myself I am small, and I garden to remind myself that I matter. Perhaps I'll have a baby, perhaps I won't. Perhaps I'll divorce, perhaps I won't.

Perhaps I'll live alone, perhaps I won't. Whatever happens, I will always return to the earth.

One dull, mild morning in late January I open the back door and go out. As I walk around the garden, peering at the beds, at the rough edge of the lawn, at the bits of rubbish that have made their way in from the city, I feel a pull. I've been away for too long.

The next day I get out of bed and dress for the garden. I cut back the last of the herbaceous plants, the crisping sticks that used to be green. I chop the fennel back to the ground, making space for the keen new fronds at the crown. I run my fingers over the new bulb shoots that are coming and admire the buds of growth. Gently, I rake over the soil, and pile bits of rubbish on the table. I scrabble around in the compost bin for the dark, crumbly goodness inside, and spread it around the beds, then put everything else in to rot fantastically. I take the edging tool and slide it between the flat, recovering brown of the lawn and the beds beyond. I am sweating beneath my layers, I have muck on my skin. This is a cleansing of winter. This is a reunion. This is the beginning of something new.

EPILOGUE

THE CAB PULLS UP BY the front gate and Matt gets out to wrangle with the suitcases. The honey garlic in the planter has grown twisted and pale, the bulbs lilting drunkenly out of crumbling soil. It must have been another dry April here.

We have been away for nearly a month, driving on highways along the ocean and through the desert, mouths agape. The stars have come out for us, the whales too. Before we left we got married on a searing day in spring. It keeps coming back to me in fragments: the sparrows hopping between the blossom in the bombed-out church where we said our vows, what it was to see our favourite people jigsaw together in one room, the smell of narcissus. We have been propped up on firm, tough endorphins for some time, now.

The cases are wheeled in and I head downstairs to the garden. I left it at the feverish start of the growing season.

We have missed the best of the tulips, the end of the daffs. In their place is abundance: rose buds creeping up the wall, five feet of sweet pea stems, a feathery fountain of fennel. The beds have swollen with green. This is a collision of foliage: the spikes of *Echinops*, the lily-pad leaves of hollyhocks, the curlicues of peony leaves. Among it all, the residue of blown-out tulips, oil-streaked with their final, withering colours.

I have sat with the question of why women grow for two years. At first, I tried to respond to it by seeking an answer; by gathering names and making requests, by sending out surveys and following the replies. It was this question that nudged me to drive across the country and take trains to its corners, sit on ferries to meet strangers and listen to them speak about their lives. As the seasons folded into one another, I collected my findings like seeds: carefully extracting the hard little beads of promise from the fluff of their protective casings, tucking them safely away into labelled envelopes. I saw connections between desire and resilience, perseverance and power, curiosity and motherhood. Often, I felt overwhelmed by the heft of these stories. It was difficult to hear them, to take them, without feeling indebted in some way.

Women grow to create life and food and beauty. Women grow to conjure substance from the scruff of land. Women grow because they see potential in the hard, rolled nuggets of clay and promise beneath the tangles of ancient shrubs.

Women grow because it makes something of nothing, because they see that broader changes can come from those they make in the small patch beneath their feet. Women grow because they are heavy with sadness, or solitude, or grief. Women grow because it is in their bones. Women grow because raising a garden can forge connective tissue to something lost. Women grow to pass on power, to honour the knowledge their foremothers have gathered for centuries. Women grow because the earth can swallow feelings that the air can't. Women grow because sometimes rage can only be mollified by digging until the sweat trickles down their backs. Women grow because in doing so they can make space – sometimes silently, sometimes by stealth – that nobody expects them to occupy. Women grow because it offers them control in a world determined to rid them of it. Women grow because they are curious, and canny, and they are compelled to.

It has taken me time to realise that when I met with these women, I was seeking guidance. That I set out not knowing how to be or to live, that I was uncomfortable with growing into a new stage of life, that I wanted permission to explain myself. I think about the stories I've heard and what I've learned from them, all the lives boiled down into cups of tea and walks around gardens. I will hold these close – I cherish them – but I know now that I have to do the work myself. It is in me, this fierce womanhood, this state of being. It is time for me to trust her.

There are ways to leave a garden. You can prepare for its abandonment for weeks, planning that season's planting months before and leaving seedlings in the hands of trusted friends. You can make arrangements for someone to come in and tend and water and deadhead and mow, so that you return to good order. Or you can simply savour what it looks like before you go, lock the back door and leave it to unfold without you.

For the first time – since that first balcony, since Treehouse, since the first proper summer in the garden – I had left it, and I had left it be. Among the sprawling, heat-wavy horizons of Arizona it was impossible to think of what was growing, or not, in our little South London plot. Among the growth are other signs of life: a fox has burrowed beneath the rose and unearthed bulbs in the sunniest corner, wrens dart in and out of the nest they've built in the clematis, snails and slugs lurk among the dying foliage. I have started this, and now it is growing into itself.

I look at the garden and I feel communion. I hear Şifa talking about how much she loves the colour green. I hear whispers of Charleston in the towering fennel. I hear Diana's wry laugh as she regales me with stories of her 'vandal mode'. I think of Sui and the ecosystem she has built by sitting back and watching. I think about Anne beneath her magnificent angelica – 'you need strength in old age' – and Hannah, sowing sunflowers. I wonder if Ali and Sarah have any buds on their roses yet. Mel's words, about feeling less alone outside, ring through the air with a whole new resonance.

Whatever tight control I once held over the garden – of the maniacal spring seed-sowing and the fretting about compost and the hurry to order plants – has drifted away. In letting it grow, I see a reflection of who I am. This garden doesn't toe the lines or have neat edges. It holds weeds and foxholes and aphids. It lights up with the hardy geraniums and hellebores of my childhood but boasts far more plants I have found for myself.

Two years have passed since I wanted to know why women gardened. Two years have passed since we moved to this plot. I have made changes in this garden, but it has changed me far more. I had wanted clear answers without realising I needed to grow to find them. To learn that womanhood is lonely because it is hard, but that loneliness is why we uphold one another. That rage is legitimate and can be power, and power makes change. That we can define our own understanding of beauty, and it doesn't matter if it's only beautiful to us. I wanted the garden to be so much, but it was only in leaving it that it became what it was always meant to be: a space for me to define my own womanhood.

This garden will grow. The flowers will open, go to

seed and wither. Much-loved perennials will be ousted by weeds, expensive bulbs will be dug up by other bits of life. So much more will survive and surprise and endure. I will do less gardening and more seeing. I will let it unfold and I will listen. I will learn from the garden as I have the women I have met.

I grab the kitchen scissors and cut a bunch of flowers for the kitchen table. The best of the tulips, the last of the narcissi, the first of the sweet peas. There will be more, yet.

FURTHER READING

Books

Gaining Ground by Joan Barfoot

Charleston: A Bloomsbury House & Garden by Quentin Bell and Virginia Nicholson

The Selected Letters of Vanessa Bell by Vanessa Bell

Grounding by Lulah Ellender

Deceived with Kindness by Angelica Garnett

A Modern Herbal by Mrs M. Grieve, edited and introduced by Mrs C.F. Leyel

Gardening Women: Their Stories from 1600 to the Present by Catherine Horwood

The Virago Book of Women Gardeners edited by Deborah Kellaway

My Garden (Book) by Jamaica Kincaid

Herlands: Exploring the Women's Land Movement in the United States by Keridwen N. Luis

Plant Form & Design by W. Midgley and A.E.V. Lilley
Sister Arts: The Erotics of Lesbian Landscapes by Lisa L. Moore
The Poetics of Gardens by Charles W. Moore, William J. Mitchell and William Turnbull, Jr.
Unearthed by Claire Ratinon
Earthed by Rebecca Schiller
Autumn by Ali Smith
Winter by Ali Smith
Spring by Ali Smith
Summer by Ali Smith
The Morville Hours by Katherine Swift
In the Garden: Essays on Nature and Growing by Various
The Gardens of the British Working Class by Margaret Willes

Articles

'How gardening helped me live with love and loss', Rhiannon Lucy Cosslett, *The Guardian*, August 2020
'I am a mother without a baby', Fiona Crack, BBC, 2018
'Not all in the mind', Diana Ross, *Hortus* No. 31, Autumn 1994

ACKNOWLEDGEMENTS

I AM INDEBTED TO THE women who gave their time, hospitality and stories for me to hear and hold. To take a stranger into your home, garden or meaningful space is a generous act; to tell them about your life is a courageous one. To those who said yes: you have enabled this book to exist, you have helped me to understand better what it is to be a woman, and you have proved that our stories are worth telling. Thank you.

Thank you to Rachel Mills, for your enthusiasm and efficiency in giving this book credence long before I believed in it. To Jo Dingley for picking it up and breathing life into it, and bolstering me with the confidence I needed to write it. Jo, I wrote this with you in mind; thank you. For Helena Gonda, who took *Why Women Grow* on as if she had been there from the beginning; your passion, your determination and your clear-eyed edit turned this book into what it is. I feel very fortunate that the book has had two fiercely talented editors.

To Claire Reiderman, thank you for your dedicated work to pull all the strings together. Gabbie Chant, thank you for your meticulous and kind-hearted copy-editing. To the rest of the Canongate team: to return with *Why Women Grow* after your careful nurturing of *Rootbound* felt like a homecoming. Thank you to Simon Thorogood, to Lucy Zhou, to Jamie Norman and to Rafi Romaya.

Thank you to those who have supported this, and me, from the beginning. To Claire Ratinon, to Maya Thomas, to Hazel Gardiner, to Hannah Murphy, to Andrew O'Brien, to Jess Bailey and to Diana Ross for your questions, your care and your championing. To my colleagues Sam Parker, Sarah McKenna and Stephen Carlick for enabling me to fit this in. Thank you to Mary Warshaw for her kind gift of 'The Bug', without which far fewer miles would have been travelled and far fewer interviews conducted. Thank you to Matt Trueman, for your patience and empowerment, always.